AF560550

INTERNATIONAL ENCYCLOPAEDIA OF BIOSCIENCE & TECHNOLOGY

INTERNATIONAL ENCYCLOPAEDIA OF BIOSCIENCE & TECHNOLOGY

INTERNATIONAL ENCYCLOPAEDIA OF
BIOSCIENCE & TECHNOLOGY

O.P. Pandey

ANMOL PUBLICATIONS PVT. LTD.
NEW DELHI - 110 002 (INDIA)

ANMOL PUBLICATIONS PVT. LTD.
H.O.: 4374/4B, Ansari Road, Daryaganj,
New Delhi-110 002 (India)
Ph.: 23278000, 23261597
B.O.: No. 1015, Ist Main Road, BSK IIIrd Stage
IIIrd Phase, IIIrd Block,
Bangalore - 560 085 (India)
Visit us at: www.anmolpublications.com

International Encyclopaedia of Bioscience & Technology

ISBN 978-81-261-3591-2

PRINTED IN INDIA

Printed at Mehra Offset Press, Delhi.

INTERNATIONAL ENCYCLOPAEDIA OF
BIOSCIENCE & TECHNOLOGY

O.P. Pandey

ANMOL PUBLICATIONS PVT. LTD.
NEW DELHI - 110 002 (INDIA)

ANMOL PUBLICATIONS PVT. LTD.
H.O.: 4374/4B, Ansari Road, Daryaganj,
New Delhi-110 002 (India)
Ph.: 23278000, 23261597
B.O.: No. 1015, Ist Main Road, BSK IIIrd Stage
IIIrd Phase, IIIrd Block,
Bangalore - 560 085 (India)
Visit us at: www.anmolpublications.com

International Encyclopaedia of Bioscience & Technology

ISBN 978-81-261-3591-2

PRINTED IN INDIA

Printed at Mehra Offset Press, Delhi.

Contents

Preface

This publication entitled "International Encyclopaedia of Bioscience & Technology" is a timely initiative to deal with the said subject in its totality. Bioscience is the study of life and is concerned with such topics, such as the characteristics, classification and behavior of organisms; how species come into existence; and the interactions they have with each other and with the natural environment. It is a science that deals with the functions or problems of living organisms. Biotech refers to the biological science, applied especially in genetic engineering and recombinant DNA technology. It deals with a set of biological techniques developed through basic research and now applied to research and product development. It is the means or way of manipulating life forms to provide desirable products for man's use. For example, beekeeping and cattle breeding could be considered to be biotechnology-related endeavors. However, the use of this technology has come to mean all parts of an industry that knowingly create, develop, and market a variety of products through the willful manipulation, on a molecular level, of life forms or utilization of knowledge pertaining to living systems. However, recombinant DNA is only one of the many techniques used to derive products from organisms, plants, and parts of both for the biotechnology industry. A list of areas covered by the term biotechnology would more properly include: recombinant DNA, plant tissue culture, rDNA or gene splicing, enzyme systems, plant breeding, meristem culture, mammalian cell culture, immunology, molecular biology, fermentation, and others. Biochemistry deals with the chemical characteristics and reactions of a particular living system or biological substance. Biopharmaceutical refers to the application of biological technology research to the development of pharmaceutical products that improve human health, animal health, and agriculture. Today, all these are a rapidly growing field of research and development that is cutting across many traditional boundaries.

This encyclopaedia titled "International Encyclopaedia of Bioscience & Technology" is really unique in terms of coverage, scope, elaboration, usefulness and futuristic perspective. It will provide readers with a user-friendly approach towards acquainting themselves with various select traditional and newly emerging fields of biosciences and related emerging technologies. This encyclopaedia is of great contemporary relevance. It will serve as a reference book on the said subject and tends to acquaint readers with both in-depth and extensive knowledge of the subject concerned. This encyclopaedia is a unique initiative to provide the most comprehensive coverage to the subject area in question. This encyclopedia mainly focuses on the following areas: Introduction to Bioscience; Modern Bioscience Branches and Areas of Studies; Origin of Life and Bioscience Timeline; Case Studies of Streams in Bioscience: Illustrations from Biophysics, Medicine & Neuroscience; Evolution, Ecology and Evolution Biology; Introduction to Technology and Bioscience Technology; Overview of Biological Technologies with Select Case Studies; Glossary etc.

—Editor

1
Introduction to Bioscience

BIOSCIENCE (BIOLOGY)

The biosciences — the branches of natural science dealing with the structure and behavior of living organisms — are as diverse as the animals, plants and microbes that make up life on Earth. The scope of the biosciences reaches across many different scales of size and of time, from the microscopic and submicroscopic levels of cells and molecules to the global scale of biological communities and ecosystems; as well as from the present through vast stretches of the past. The bioscience disciplines are highly allied and interdependent, and some, such as biophysics and biomedical engineering, transcend the traditional boundaries of the physical sciences and of technology (the so-called "applied sciences"). The descriptions of various bioscience disciplines presented here are intended to be brief introductions only — snapshots, not formal portraits. Each contains a concise explanation of the field, a representative list of subfields within the discipline, some career options and several links to Web sites for additional information. We encourage you to use the information and links to start your own investigation into the "sciences of life."

Biology as a unified science was first developed in the nineteenth century, as scientists discovered that all living things shared certain fundamental characteristics and were best studied as a whole. Over a million papers are published annually in a wide array of biology and medicine journals, and biology is a standard subject of instruction at schools and universities around the world.

Biology is the study of life. It contains such topics as classifying the various forms of organisms, how species come into existence, and the interactions they have with each other and with the natural environment.

Biology encompasses a broad spectrum of academic fields that are often viewed as independent disciplines. However, together they address phenomena related to living organisms (biological phenomena) over a wide range of disciplines, many of which, for example, botany, zoology, and medicine are considered ancient fields of study.

Biology has become such a vast research enterprise that it is not generally regarded as a single discipline, but a number do assist in understanding the genetic variation of a population; and physiology borrows extensively from cell biology in describing the function of organ systems. Ethology and comparative psychology extend biology to the analysis of animal behavior and mental characteristics, whilst Evolutionary psychology proposes that the field of psychology, including in regard to humans, is a branch of biology.

As such a vast field, biology is divided into a number of disciplines. The old divisions by type of organism remains with subjects such as botany encompassing the study of plants, zoology with the study of animals, and microbiology as the study of microorganisms.

The field may also be divided based on the scale at which it is studied: biochemistry examines the fundamental chemistry of life; cellular biology examines the basic building block of all life, the cell; Physiology examines the mechanical and physical functions of an organism; and ecology examines how various organisms interrelate. Applied fields of biology such as medicine are more complex and involve many specialized sub-disciplines.

Structure of Life

Molecular biology is the study of biology at a molecular level. This field overlaps with other areas of biology, particularly with genetics and biochemistry. Molecular biology chiefly concerns itself with understanding the interactions between the various systems of a cell, including the interrelationship of DNA, RNA, and protein synthesis and learning how these interactions are regulated. Cell biology studies the physiological properties of cells, as well as their behaviors, interactions, and environment. This is done both on a microscopic and molecular level. Cell biology researches both single-celled organisms like bacteria and specialized cells in multicellular organisms like humans.

Understanding cell composition and how they function is fundamental to all of the biological sciences. Appreciating the similarities and differences between cell types is particularly important in the fields of cell and molecular biology. These fundamental similarities and differences provide a unifying theme, allowing the principles learned from studying one cell type to be extrapolated and generalized to other cell types.

Genetics is the science of genes, heredity, and the variation of organisms. Genes encode the information necessary for synthesizing proteins, which in turn play a large role in influencing (though, in many instances, not completely determining) the final phenotype of the organism. In modern research, genetics provides important tools in the investigation of the function of a particular gene, or the analysis of genetic interactions. Within organisms, genetic information generally is carried in chromosomes, where it is represented in the chemical structure of particular DNA molecules.

Developmental biology studies the process by which organisms grow and develop. Originating in embryology, modern developmental biology studies the genetic control of cell growth, differentiation, and "morphogenesis," which is the process that gives rise to tissues, organs, and anatomy. Model organisms for developmental biology include the round worm *Caenorhabditis elegans*, the fruit fly *Drosophila melanogaster*, the zebrafish *Brachydanio rerio*, the mouse *Mus musculus*, and the weed *Arabidopsis thaliana*.

Common Descent

Up into the 19th century, it was commonly believed that life forms could appear spontaneously under certain conditions. This misconception was challenged by William Harvey's diction that "all life [is] from [an] egg" (from the Latin "Omne vivum ex ovo"), a foundational concept of modern biology. It simply means that there is an unbroken continuity of life from its initial origin to the present time.

A group of organisms shares a common descent if they share a common ancestor. All organisms on the Earth have been and are descended from a common ancestor or an ancestral gene pool. This last universal common ancestor of all organisms is believed to have appeared about 3.5 billion years ago. Biologists generally regard the universality of the genetic code as definitive evidence in favor of the theory of universal common descent (UCD) for all bacteria, archaea, and eukaryotes.

Diversity and Evolution of Organisms

Biodiversity, Botany, Zoology is concerned with the origin and descent of species, as well as their change over time, and includes scientists from many taxonomically-oriented disciplines. For example, it generally involves scientists who have special training in particular organisms such as mammalogy, ornithology, or herpetology, but use those organisms as systems to answer general questions about evolution. Evolutionary biology is mainly based on paleontology, which uses the fossil record to answer questions about the mode and tempo of evolution, as well as the developments in areas such as population genetics and evolutionary theory. In the 1990s, developmental biology re-entered evolutionary biology from its initial exclusion from the modern synthesis through the study of evolutionary developmental biology. Related fields which are often considered part of evolutionary biology are phylogenetics, systematics, and taxonomy.

The two major traditional taxonomically-oriented disciplines are botany and zoology. Botany is the scientific study of plants. Botany covers a wide range of scientific disciplines that study the growth, reproduction, metabolism, development, diseases, and evolution of plant life.

Zoology involves the study of animals, including the study of their physiology within the fields of anatomy and embryology. The common genetic and developmental mechanisms of animals and plants is studied in molecular biology, molecular genetics, and developmental biology. The ecology of animals is covered under behavioral ecology and other fields.

Physiology of Organisms

Physiology studies the mechanical, physical, and biochemical processes of living organisms by attempting to understand how all of the structures function as a whole. The theme of "structure to function" is central to biology. Physiological studies have traditionally been divided into plant physiology and animal physiology, but the principles of physiology are universal, no matter what particular organism is being studied. For example, what is learned about the physiology of yeast cells can also apply to human cells. The field of animal physiology extends the tools and methods of human physiology to non-human species. Plant physiology also borrows techniques from both fields.

Anatomy is an important branch of physiology and considers how organ systems in animals, such as the nervous, immune, endocrine, respiratory, and circulatory systems, function and interact. The study of these systems is shared with medically oriented disciplines such as neurology and immunology.

Interactions of Organisms

Ecology studies the distribution and abundance of living organisms, and the interactions between organisms and their environment. The environment of an organism includes both its habitat, which can be described as the sum of local abiotic factors such as climate and ecology, as well as the other the organisms that share its habitat. Ecological systems are studied at several different levels, from individuals and populations to ecosystems and the biosphere. As can be surmised, ecology is a science that draws on several disciplines. Ethology studies animal behavior (particularly of social animals such as primates and canids), and is sometimes considered a branch of zoology. Ethologists have been particularly concerned with the evolution of behavior and the understanding of behavior in terms of the theory of natural selection. In one sense, the first modern ethologist was Charles Darwin, whose book "The Expression of the Emotions in Man and Animals" influenced many ethologists.

Biogeography studies the spatial distribution of organisms on the Earth, focusing on topics like plate tectonics, climate change, dispersal and migration, and cladistics.

Every living thing interacts with other organisms and its environment. One reason that biological systems can be difficult to study is that so many different interactions with other organisms and the environment are possible, even on the smallest of scales. A microscopic bacterium responding to a local sugar gradient is responding to its environment as much as a lion is responding to its environment when it searches for food in the African savannah. For any given species, behaviors can be co-operative, aggressive, parasitic or symbiotic. Matters become more complex when two or more different species interact in an ecosystem. Studies of this type are the province of ecology.

Although the concept of *biology* as a single coherent field arose in the 19th century, the biological sciences emerged from traditions of medicine and natural history reaching back to Galen and Aristotle in ancient Greece. During the Renaissance and early modern period, biological thought was revolutionized by a renewed interest in empiricism and the discovery of many novel organisms. Prominent in this movement were Vesalius and Harvey, who used experimentation and careful observation in physiology, and naturalists such as Linnaeus and Buffon who began to classify the diversity of life and the fossil record, as well as the development and behavior of organisms. Microscopy revealed the previously unknown world of microorganisms, laying the groundwork for cell theory. The growing importance of natural theology, partly a response to the rise of mechanical philosophy, encouraged the growth of natural history (though it entrenched the argument from design).

Over the 18th and 19th centuries, biological sciences such as botany and zoology became increasingly professional scientific disciplines. Lavoisier and other physical scientists began to connect the animate and inanimate worlds through physics and chemistry. Explorer-naturalists such as Alexander von Humboldt investigated the interaction between organisms and their environment, and the ways this relationship depends on geography—laying the foundations for biogeography, ecology and ethology. Naturalists began to reject essentialism and consider the importance of extinction and the mutability of species. Cell theory provided a new perspective on the fundamental basis of life. These developments, as well as the results from embryology and paleontology, were synthesized in Charles Darwin's theory of evolution by natural selection. The end of the 19th century saw the fall of spontaneous generation and the rise of the germ theory of disease, though the mechanism of inheritance remained a mystery.

In the early 20th century, the rediscovery of Mendel's work led to the rapid development of genetics by Thomas Hunt Morgan and his students, and by the 1930s the combination of population genetics and natural selection in the "neo-Darwinian synthesis". New disciplines developed rapidly, especially after Watson and Crick proposed the structure of DNA. Following the establishment of the Central Dogma and the cracking of the genetic code, biology was largely split between *organismal biology*—the fields that deal with whole organisms and groups of organisms—and the fields related to *cellular and molecular biology*. By the late 20th century, new fields like genomics and proteomics were reversing this trend, with organismal biologists using molecular techniques, and molecular and cell biologists investigating the interplay between genes and the environment, as well as the genetics of natural populations of organisms.

Taxonomy

Classification is the province of the disciplines of systematics and taxonomy. Taxonomy places organisms in groups called taxa, while systematics seeks to define their relationships with each other. This classification technique has evolved to reflect advances in cladistics and genetics, shifting the focus from physical similarities and shared characteristics to phylogenetics.

Traditionally, living things have been divided into five kingdoms:

Monera — Protista — Fungi — Plantae — Animalia

However, many scientists now consider this five-kingdom system to be outdated. Modern alternative classification systems generally begin with the three-domain system:

Archaea (originally Archaebacteria) — Bacteria (originally Eubacteria) — Eukaryota

The scientific name of an organism is obtained from its genus and species. For example, humans would be listed as *Homo sapiens*. *Homo* would be the genus and *sapiens* is the species. Whenever writing the scientific name of an organism, it is proper to capitalize the first letter in the genus and put all of the species in lowercase; in addition the entire term would be put in italics or underlined. The term used for classification is called taxonomy. There is also a series of intracellular parasites that are progressively "less alive" in terms of metabolic activity:

Viruses — Viroids — Prions

The dominant classification system is called Linnaean taxonomy, which includes ranks and binomial nomenclature. How organisms are named is governed by international agreements such as the International Code of Botanical Nomenclature

(ICBN), the International Code of Zoological Nomenclature (ICZN), and the International Code of Nomenclature of Bacteria (ICNB). A fourth Draft BioCode was published in 1997 in an attempt to standardize naming in these three areas, but it has yet to be formally adopted. The Virus cInternational Code of Virus Classification and Nomenclature (ICVCN) remains outside the BioCode.

Foundations

Biology is a branch of science that characterizes and investigates living organisms utilizing the scientific method. There are four broad unifying principles of biology:

1. *Cell theory.* All living organisms are composed of at least one cell and the cell is the basic unit of function in all organisms. In addition, the chemical composition of all cells in all organisms is similar, and emerge from preexisting cells through cell division or mitosis.
2. *Evolution.* Through natural selection or genetic drift, a population's inherited traits change from generation to generation.
3. *Gene theory.* A living organism's traits are encoded into DNA that is the fundamental component of genes. In addition, genes transfer an organism's traits from one generation to the next.
4. *Homeostasis.* The physiological processes that allow an organism to maintain its internal environment notwithstanding its external environment.

1. Cell Theory

Essentially, cell theory states that all living things are composed of one or more cells (and products of those cells, for example, plasma, saliva, and other substances). Furthermore, those cells arise from other cells through cellular division, all cells in all organisms are similar in chemical composition, and that energy flow or metabolism occurs within the cells.

2. Evolution

A central organizing concept in biology is that all life has a common origin and has changed and developed through the process of the theory of evolution. This has led to the striking similarity of units and processes discussed in the previous section. Charles Darwin established evolution as a viable theory by articulating its driving force, natural selection (Alfred Russel Wallace is recognized as the co-discoverer of this concept). Darwin theorized that species and breeds developed through the processes of natural selection as well as by artificial selection or selective breeding.Genetic drift was embraced as an additional mechanism of evolutionary development in the modern synthesis of the theory.

The evolutionary history of a species— which describes the characteristics of the various species from which it descended— together with its genealogical relationship to every other species is called its phylogeny. Widely varied approaches to biology generate information about phylogeny. These include the comparisons of DNA sequences conducted within molecular biology or genomics, and comparisons of fossils or other records of ancient organisms in paleontology. Biologists organize and analyze evolutionary relationships through various methods, including phylogenetics, phenetics, and cladistics (The major events in the evolution of life, as biologists currently understand them, are summarized on this evolutionary timeline).

Ever since its articulation by Darwin and Wallace, the theory of evolution by natural selection has come under attack by people who disagree with scientific findings or interpretations regarding the origins and diversity of life, generally favoring instead religious explanations. See Creation-evolution controversy for more information.

Up into the 19th century, it was commonly believed that life forms could appear spontaneously under certain conditions. This misconception was challenged by William Harvey's diction that "all life [is] from [an] egg" (from the Latin "Omne vivum ex ovo"), a foundational concept of modern biology. It simply means that there is an unbroken continuity of life from its initial origin to the present time.

A group of organisms shares a common descent if they share a common ancestor. All organisms on the Earth have been and are descended from a common ancestor or an ancestral gene pool. This last universal common ancestor of all organisms is believed to have appeared about 3.5 billion years ago. Biologists generally regard the universality of the genetic code as definitive evidence in favor of the theory of universal common descent (UCD) for all bacteria, archaea, and eukaryotes.

3. Gene Theory

While organisms may vary immensely in appearance, habitat, and behaviour it is a central principle of biology that all life shares certain universal fundamentals. A key feature is reproduction or replication. The entity being replicated, the replicator, in the past was considered to be the organism during the time of Darwin, but since the 1970s increasingly reduced to the scale of molecules. All known life has a carbon-based biochemistry, carbon is the fundamental building block of the molecules that make up all known living things. Similarly water is the basic solvent for all known living organisms. While all these things are true of all organisms observed on Earth, in theory alternative forms of life could exist and some scientists do look at alternative biochemistry.

All terrestrial organisms use DNA and RNA-based genetic mechanisms to hold genetic information. Another universal principle is that all observed organisms with the exception of viruses are made of cells. Similarly, all organisms share common developmental processes.

4. Homeostasis

Homeostasis is the ability of an open system to regulate its internal environment to maintain a stable condition by means of multiple dynamic equilibrium adjustments controlled by interrelated regulation mechanisms. All living organisms, whether unicellular or multicellular, exhibit homeostasis. Homeostasis manifests itself at the cellular level through the maintenance of a stable internal acidity (pH); at the organismic level, warm-blooded animals maintain a constant internal body temperature; and at the level of the ecosystem, as when atmospheric carbon dioxide levels rise and plants are theoretically able to grow healthier and remove more of the gas from the atmosphere. Tissues and organs can also maintain homeostasis.

GENE

A gene is a locatable region of genomic sequence, corresponding to a unit of inheritance, which is associated with regulatory regions, transcribed regions and/or other functional sequence regions. The physical development and phenotype of organisms can be thought of as a product of genes interacting with each other and with the environment. A concise definition of gene taking into account complex patterns of regulation and transcription, genic conservation and non-coding RNA genes, has been proposed by Gerstein et al. "A gene is a union of genomic sequences encoding a coherent set of potentially overlapping functional products".

In cells, genes consist of a long strand of DNA that contains a promoter, which controls the activity of a gene, and coding and non-coding sequence. Coding sequence determines what the gene produces, while non-coding sequence can regulate the conditions of gene expression. When a gene is active, the coding and non-coding sequence is copied in a process called transcription, producing an RNA copy of the gene's information. This RNA can then direct the synthesis of proteins via the genetic code. However, RNAs can also be used directly, for example as part of the ribosome. These molecules resulting from gene expression, whether RNA or protein, are known as gene products.

Genes often contain regions that do not encode products, but regulate gene expression. The genes of eukaryotic organisms can contain regions called introns that are removed from the messenger RNA in a process called splicing. The regions encoding gene products are called exons. In eukaryotes, a single gene can encode multiple proteins, which are produced through the creation of different arrangements of exons through *alternative splicing*. In prokaryotes (bacteria and archaea), introns are less common and genes often contain a single uninterrupted stretch of DNA, called a *cistron*, that codes for a product. Prokaryotic genes are often arranged in groups called operons with promoter and operator sequences that regulate transcription of a single long RNA. This RNA contains multiple coding sequences. Each coding sequence is preceded by a Shine-Dalgarno sequence that ribosomes recognize.

The total set of genes in an organism is known as its genome. An organism's genome size is generally lower in prokaryotes, both in number of base pairs and number of genes, than even single-celled eukaryotes. However, there is no clear relationship between genome sizes and complexity in eukaryotic organisms. One of the largest known genomes belongs to the single-celled amoeba *Amoeba dubia*, with over 670 billion base pairs, some 200 times larger than the human genome. The estimated number of genes in the human genome has been repeatedly revised downward since the completion of the Human Genome Project; current estimates place the human genome at just under 3 billion base pairs and about 20,000–25,000 genes.. A recent *Science* article gives a final number of 20,488, with perhaps 100 more yet to be discovered. The gene density of a genome is a measure of the number of genes per million base pairs (called a megabase, Mb); prokaryotic genomes have much higher gene densities than eukaryotes. The gene density of the human genome is roughly 12–15 genes/Mb.

History

The existence of genes was first suggested by Gregor Mendel (1822-1884), who, in the 1860s, studied inheritance in pea plants and hypothesized a factor that conveys traits from parent to offspring. He spent over 10 years of his life on one experiment. Although he did not use the term *gene*, he explained his results in terms of inherited characteristics. Mendel was also the first to hypothesize independent assortment, the distinction between dominant and recessive traits, the distinction between a heterozygote and homozygote, and the difference between what would later be described as genotype and phenotype. Mendel's concept was given a name by Hugo de Vries in 1889, who, at that time probably unaware of Mendel's work, in his book *Intracellular Pangenesis* coined the term "pangen" for "the smallest particle [representing] one hereditary characteristic". Wilhelm Johannsen abbreviated this term to "gene" ("gen" in Danish and German) two decades later.

In the early 1900s, Mendel's work received renewed attention from scientists. In 1910, Thomas Hunt Morgan showed that genes reside on specific chromosomes. He later showed that genes occupy specific locations on the chromosome. With this knowledge, Morgan and his students began the first chromosomal map of the fruit fly *Drosophila*. In 1928, Frederick Griffith showed that genes could be transferred. In what is now known as Griffith's experiment, injections into a mouse of a deadly strain of bacteria that had been heat-killed transferred genetic information to a safe strain of the same bacteria, killing the mouse.

In 1941, George Wells Beadle and Edward Lawrie Tatum showed that mutations in genes caused errors in certain steps in metabolic pathways. This showed that specific genes code for specific proteins, leading to the "one gene, one enzyme" hypothesis. Oswald Avery, Collin Macleod, and Maclyn McCarty showed in 1944 that DNA holds the gene's information. In 1953, James D. Watson and Francis Crick demonstrated the molecular structure of DNA. Together, these discoveries established the central dogma of molecular biology, which states that proteins are translated from RNA which is transcribed from DNA. This dogma has since been shown to have exceptions, such as reverse transcription in retroviruses.

In 1972, Walter Fiers and his team at the Laboratory of Molecular Biology of the University of Ghent (Ghent, Belgium) were the first to determine the sequence of a gene: the gene for Bacteriophage MS2 coat protein. Richard J. Roberts and Phillip Sharp discovered in 1977 that genes can be split into segments. This leads to the idea that one gene can make several proteins. Recently (as of 2003-2006), biological results let the notion of gene appear more slippery. In particular, genes do not seem to sit side by side on DNA like discrete beads. Instead, regions of the DNA producing distinct proteins may overlap, so that the idea emerges that "genes are one long continuum".

Mendelian Inheritance and Classical Genetics

Darwin used the term Gemmule to describe a microscopic unit of inheritance, and what would later become known as Chromosomes had been observed separating out during cell division by Wilhelm Hofmeister as early as 1848. The idea that chromosomes were the carriers of inheritance was expressed in 1883 by Wilhelm Roux. The modern conception of the gene originated with work by Gregor Mendel, a 19th century Augustinian monk who systematically studied heredity in pea plants. Mendel's work was the first to illustrate particulate inheritance, or the theory that inherited traits are passed from one generation to the next in discrete units that interact in well-defined ways. Danish botanist Wilhelm Johannsen coined the word "gene" in 1909 to describe these fundamental physical and functional units of heredity, while the related word genetics was first used by William Bateson in 1905. The word was derived from Hugo De Vries' 1889 term *pangen* for the same concept, itself a derivative of the word *pangenesis* coined by Darwin (1868). The word pangenesis is made from the Greek words *pan* (a prefix meaning "whole", "encompassing") and *genesis* ("birth") or *genos* ("origin").

According to the theory of Mendelian inheritance, variations in phenotype - the observable physical and behavioral characteristics of an organism - are due to variations in genotype, or the organism's particular set of genes, each of which specifies a particular trait. Different genes for the same trait, which give rise to different phenotypes, are known as alleles. Organisms such as the pea plants Mendel worked on, along with many plants and animals, have two alleles for each trait, one inherited from each parent. Alleles may be dominant or recessive; dominant alleles give rise to their corresponding phenotypes when paired with any other allele for the same trait, while recessive alleles give rise to their corresponding phenotype only when paired with another copy of the same allele. For example, if the allele specifying tall stems in pea plants is dominant over the allele specifying short stems, then pea plants that inherit one tall allele from one parent and one short allele from the other parent will also have tall stems. Mendel's work found that alleles assort independently in the production of gametes, or germ cells, ensuring variation in the next generation.

Prior to Mendel's work, the dominant theory of heredity was one of blending inheritance, which proposes that the traits of the parents blend or mix in a smooth, continuous gradient in the offspring. Although Mendel's work was largely unrecognized after its first publication in 1866, it was rediscovered in 1900 by three European scientists, Hugo de Vries, Carl Correns, and Erich von Tschermak, who had reached similar conclusions from their own research. However, these scientists were not yet aware of the identity of the 'discrete units' on which genetic material resides.

A series of subsequent discoveries led to the realization decades later that chromosomes within cells are the carriers of genetic material, and that they are made of DNA (deoxyribonucleic acid), a polymeric molecule found in all cells on which the 'discrete units' of Mendelian inheritance are encoded. The modern study of genetics at the level of DNA is known as molecular genetics and the synthesis of molecular genetics with traditional Darwinian evolution is known as the modern evolutionary synthesis.

Physical Definitions

The vast majority of living organisms encode their genes in long strands of DNA. DNA consists of a chain made from four types of nucleotide subunits: adenosine, cytidine, guanosine, and thymidine. Each nucleotide subunit consists of three components: a phosphate group, a deoxyribose sugar ring, and a nucleobase. Thus, nucleotides in DNA or RNA are typically called 'bases'; consequently they are commonly referred to simply by their purine or pyrimidine original base components adenine, cytosine, guanine, thymine. Adenine and guanine are purines and cytosine and thymine are pyrimidines. The most common form of DNA in a cell is in a double helix structure, in which two individual DNA strands twist around each other in a right-handed spiral. In this structure, the base pairing rules specify that guanine pairs with cytosine and adenine pairs with thymine (each pair contains one purine and one pyrimidine). The base pairing between guanine and cytosine forms three hydrogen bonds, while the base pairing between adenine and thymine forms two hydrogen bonds. The two strands in a double helix must therefore be *complementary*, that is, their bases must align such that the adenines of one strand are paired with the thymines of the other strand, and so on.

Due to the chemical composition of the pentose residues of the bases, DNA strands have directionality. One end of a DNA polymer contains an exposed hydroxyl group on the deoxyribose, this is known as the 3' end of the molecule. The other end contains an exposed phosphate group, this is the 5' end. The directionality of DNA is vitally important to many cellular processes, since double helices are necessarily directional (a strand running 5'-3' pairs with a complementary strand running 3'-5') and processes such as DNA replication occur in only one direction. All nucleic acid synthesis in a cell occurs in the 5'-3' direction, because new monomers are added via a dehydration reaction that uses the exposed 3' hydroxyl as a nucleophile.

The expression of genes encoded in DNA begins by transcribing the gene into RNA, a second type of nucleic acid that is very similar to DNA, but whose monomers contain the sugar ribose rather than deoxyribose. RNA also contains the base uracil in place of thymine. RNA molecules are less stable than DNA and are typically single-stranded. Genes that encode proteins are composed of a series of three-nucleotide sequences called codons, which serve as the "words" in the genetic "language". The genetic code specifies the correspondence during protein translation between codons and amino acids. The genetic code is nearly the same for all known organisms.

RNA Genes

In some cases, RNA is an intermediate product in the process of manufacturing proteins from genes. However, for other gene sequences, the RNA molecules are the actual functional products. For example, RNAs known as ribozymes are capable of enzymatic function, and miRNAs have a regulatory role. The DNA sequences from which such RNAs are transcribed are known as non-coding DNA, or RNA genes.

Some viruses store their entire genomes in the form of RNA, and contain no DNA at all. Because they use RNA to store genes, their cellular hosts may synthesize their proteins as soon as they are infected and without the delay in waiting for transcription. On the other hand, RNA retroviruses, such as HIV, require the reverse transcription of their genome from RNA into DNA before their proteins can be synthesized. In 2006, French researchers came across a puzzling example of RNA-mediated inheritance in mouse. Mice with a loss-of-function mutation in the gene Kit have white tails. Offspring of these mutants can have white tails despite having only normal Kit genes. The research team traced this effect back to mutated Kit RNA. While RNA is common as genetic storage material in viruses, in mammals in particular RNA inheritance has been observed very rarely.

Functional Structure of a Gene

All genes have regulatory regions in addition to regions that explicitly code for a protein or RNA product. A universal regulatory region shared by all genes is known as the promoter, which provides a position that is recognized by the transcription machinery when a gene is about to be transcribed and expressed. Although promoter regions have a consensus sequence that is the most common sequence at this position, some genes have "strong" promoters that bind the transcription machinery well, and others have "weak" promoters that bind poorly. These weak promoters usually permit a lower rate of transcription than the strong promoters, because the transcription machinery binds to them and initiates transcription less frequently. Other possible regulatory regions include enhancers, which can compensate for a weak promoter. Most regulatory regions are "upstream" — that is, before or toward the 5' end of the transcription initiation site. Eukaryotic promoter regions are much more complex and difficult to identify than prokaryotic promoters.

Many prokaryotic genes are organized into operons, or groups of genes whose products have related functions and which are transcribed as a unit. By contrast, eukaryotic genes are transcribed only one at a time, but may include long stretches of DNA called introns which are transcribed but never translated into protein (they are spliced out before translation). Splicing can also occur in prokaryotic genes, but is less common than in eukaryotes.

Chromosomes

The total complement of genes in an organism or cell is known as its genome, which may be stored on one or more chromosomes; the region of the chromosome at which a particular gene is located is called its locus. A chromosome consists of a single, very long DNA helix on which thousands of genes are encoded. Prokaryotes - bacteria and archaea - typically store their genomes on a single large, circular chromosome, sometimes supplemented by additional small circles of DNA called plasmids, which usually encode only a few genes and are easily transferable between individuals. For example, the genes for antibiotic resistance are usually encoded on bacterial plasmids and can be passed between individual cells, even those of different species, via horizontal gene transfer. Although some simple eukaryotes also possess plasmids with small numbers of genes, the majority of eukaryotic genes are stored on multiple linear chromosomes, which are packed within the nucleus in complex with storage proteins called histones. The manner in which DNA is stored on the histone, as well as chemical modifications of the histone itself, are regulatory mechanisms governing whether a particular region of DNA is accessible for gene expression. The ends of eukaryotic chromosomes are capped by long stretches of repetitive sequences called telomeres, which do not code for any gene product but are present to prevent degradation of coding and regulatory regions during DNA replication. The length of the telomeres tends to decrease each time the genome is replicated in preparation for cell division; the loss of telomeres has been proposed as an explanation for cellular senescence, or the loss of the ability to divide, and by extension for the aging process in organisms.

While the chromosomes of prokaryotes are relatively gene-dense, those of eukaryotes often contain so-called "junk DNA", or regions of DNA that serve no obvious function. Simple single-celled eukaryotes have relatively small amounts of such DNA, while the genomes of complex multicellular organisms, including humans, contain an absolute majority of DNA without an identified function. However it now appears that, although protein-coding DNA makes up barely 2% of the human genome, about 80% of the bases in the genome may be being expressed, so the term "junk DNA" may be a misnomer.

Gene Expression

In all organisms, there are two major steps separating a protein-coding gene from its protein: first, the DNA on which the gene resides must be *transcribed* from DNA to messenger RNA (mRNA), and second, it must be *translated* from mRNA to protein. RNA-coding genes must still go through the first step, but are not translated into protein. The process of producing a biologically functional molecule of either RNA or protein is called gene expression, and the resulting molecule itself is called a gene product.

Genetic Code

The genetic code is the set of rules by which a gene is translated into a functional protein. Each gene consists of a specific sequence of nucleotides encoded in a DNA (or sometimes RNA) strand; a correspondence between nucleotides, the basic building blocks of genetic material, and amino acids, the basic building blocks of proteins, must be established for genes to be successfully translated into functional proteins. Sets of three nucleotides, known as codons, each correspond to a specific amino acid or to a signal; three codons are known as "stop codons" and, instead of specifying a new amino acid, alert the translation machinery that the end of the gene has been reached. There are 64 possible codons (four possible nucleotides at each of three positions, hence 43 possible codons) and only 20 standard amino acids; hence the code is redundant and multiple codons can specify the same amino acid. The correspondence between codons and amino acids is nearly universal among all known living organisms.

Transcription

The process of genetic transcription produces a single-stranded RNA molecule known as messenger RNA, whose nucleotide sequence is complementary to the DNA from which it was transcribed. The DNA strand whose sequence matches that of the RNA is known as the coding strand and the strand from which the RNA was synthesized is the template strand. Transcription is performed by an enzyme called an RNA polymerase, which reads the template strand in the 3' to 5' direction and synthesizes the RNA from 5' to 3'. To initiate transcription,the polymerase first recognizes and binds a promoter region of the gene. Thus a major mechanism of gene regulation is the blocking or sequestering of the promoter region, either by tight binding by repressor molecules that physically block the polymerase, or by organizing the DNA so that the promoter region is not accessible.

In prokaryotes, transcription occurs in the cytoplasm; for very long transcripts, translation may begin at the 5' end of the RNA while the 3' end is still being transcribed. In eukaryotes, transcription necessarily occurs in the nucleus, where the cell's DNA is sequestered; the RNA molecule produced by the polymerase is known as the primary transcript and must undergo post-transcriptional modifications before being exported to the cytoplasm for translation. The splicing of introns present within the transcribed region is a modification unique to eukaryotes; alternative splicing mechanisms can result in mature transcripts from the same gene having different sequences and thus coding for different proteins. This is a major form of regulation in eukaryotic cells.

Translation

Translation is the process by which a mature mRNA molecule is used as a template for synthesizing a new protein. Translation is carried out by ribosomes, large complexes of RNA and protein responsible for carrying out the chemical reactions to add new amino acids to a growing polypeptide chain by the formation of peptide bonds. The genetic code is read three nucleotides at a time, in units called codons, via interactions with specialized RNA molecules called transfer RNA (tRNA). Each tRNA has three unpaired bases known as the anticodon that are complementary to the codon it reads; the tRNA is also covalently attached to the amino acid specified by the complementary codon. When the tRNA binds to its complementary codon in an mRNA strand, the ribosome ligates its amino acid cargo to the new polypeptide chain, which is synthesized from amino terminus to carboxyl terminus. During and after its synthesis, the new protein must fold to its active three-dimensional structure before it can carry out its cellular function.

DNA Replication and Inheritance

The growth, development, and reproduction of organisms relies on cell division, or the process by which a single cell divides into two usually identical daughter cells. This requires first making a duplicate copy of every gene in the genome in a process called DNA replication. The copies are made by specialized enzymes known as DNA polymerases, which "read" one strand of the double-helical DNA, known as the template strand, and synthesize a new complementary strand. Because the DNA double helix is held together by base pairing, the sequence of one strand completely specifies the sequence of its complement; hence only one strand needs to be read by the enzyme to produce a faithful copy. The process of DNA replication is semiconservative; that is, the copy of the genome inherited by each daughter cell contains one original and one newly synthesized strand of DNA.

After DNA replication is complete, the cell must physically separate the two copies of the genome and divide into two distinct membrane-bound cells. In prokaryotes - bacteria and archaea - this usually occurs via a relatively simple process called binary fission, in which each circular genome attaches to the cell membrane and is separated into the daughter cells as the membrane invaginates to split the cytoplasm into two membrane-bound portions. Binary fission is extremely fast compared to the rates of cell division in eukaryotes. Eukaryotic cell division is a more complex process known as the cell cycle; DNA replication occurs during a phase of this cycle known as S phase, while the process of segregating chromosomes and splitting the cytoplasm occurs during M phase. In many single-celled eukaryotes such as yeast, reproduction by budding is common, which results in asymmetrical portions of cytoplasm in the two daughter cells.

Molecular Inheritance

The duplication and transmission of genetic material from one generation of cells to the next is the basis for molecular inheritance, and the link between the classical and molecular pictures of genes. Organisms inherit the characteristics of their parents because the cells of the offspring contain copies of the genes in their parents' cells. In asexually reproducing organisms, the offspring will be a genetic copy or clone of the parent organism. In sexually reproducing organisms, a specialized form of cell division called meiosis produces cells called gametes or germ cells that are haploid, or contain only one copy of each gene. The gametes produced by females are called eggs or ova, and those produced by males are called sperm. Two gametes fuse to form a fertilized egg, a single cell that once again has a diploid number of genes - each with one copy from the mother and one copy from the father.

During the process of meiotic cell division, an event called genetic recombination or *crossing-over* can sometimes occur, in which a length of DNA on one chromatid is swapped with a length of DNA on the corresponding sister chromatid. This has no effect if the alleles on the chromatids are the same, but results in reassortment of otherwise linked alleles if they are different. The Mendelian principle of independent assortment asserts that each of a parent's two genes for each trait will sort independently into gametes; which allele an organism inherits for one trait is unrelated to which allele it inherits for another trait. This is in fact only true for genes that do not reside on the same chromosome, or are located very far from one another on the same chromosome. The closer two genes lie on the same chromosome, the more closely they will be associated in gametes and the more often they will appear together; genes that are very close are essentially never separated because it is extremely unlikely that a crossover point will occur between them. This is known as genetic linkage.

Mutation

DNA replication is for the most part extremely accurate, with an error rate per site of around 10-6 to 10-10 in eukaryotes. Rare, spontaneous alterations in the base sequence of a particular gene arise from a number of sources, such as errors in DNA replication and the aftermath of DNA damage. These errors are called mutations. The cell contains many DNA repair mechanisms for preventing mutations and maintaining the integrity of the genome; however, in some cases — such as breaks in both DNA strands of a chromosome — repairing the physical damage to the molecule is a higher priority than producing an exact copy. Due to the degeneracy of the genetic code, some mutations in protein-coding genes are *silent*, or produce no change in the amino acid sequence of the protein for which they code; for example, the codons UCU and UUC both code for serine, so the U"!C mutation has no effect on the protein. Mutations that do have phenotypic effects are most often neutral or deleterious to the organism, but sometimes they confer benefits to the organism's fitness.

Mutations propagated to the next generation lead to variations within a species' population. Variants of a single gene are known as alleles, and differences in alleles may give rise to differences in traits. Although it is rare for the variants in a single gene to have clearly distinguishable phenotypic effects, certain well-defined traits are in fact controlled by single genetic loci. A gene's most common allele is called the wild type allele, and rare alleles are called mutants. However, this does not imply that the wild-type allele is the ancestor from which the mutants are descended.

The Genome

Chromosomal Organization

The total complement of genes in an organism or cell is known as its genome. In prokaryotes, the vast majority of genes are located on a single chromosome of circular DNA, while eukaryotes usually possess multiple individual linear DNA helices packed into dense DNA-protein complexes called chromosomes.

Extrachromosomal DNA is present in many prokaryotes and some simple eukaryotes as small, circular pieces of DNA called plasmids, which usually contain only a few genes each. Generally, regulatory regions and junk DNA are considered to be part of an organism's genome, but structural regions such as telomeres are not. The location (or locus) of a gene and the chromosome on which it is situated is in a sense arbitrary. Genes that appear together on the chromosomes of one species, such as humans, may appear on separate chromosomes in another species, such as mice. Two genes positioned near one another on a chromosome may encode proteins that figure in the same cellular process or in completely unrelated processes. As an example of the former, many of the genes involved in spermatogenesis reside together on the Y chromosome.

Many species carry more than one copy of their genome within each of their somatic cells. Cells or organisms with only one copy of each gene are called haploid; those with two copies are called diploid; and those with more than two copies are called polyploid. When more than one copy is present, the two copies are not necessarily identical; in sexually reproducing organisms, one copy is normally inherited from each parent. The copies may contain distinct DNA sequences encoding distinct alleles.

Composition of the Genome

Typical numbers of genes and size of genomes vary widely among organisms, even those that are fairly closely evolutionarily related. Although it was believed before the completion of the Human Genome Project that the human genome would contain many more genes than simpler animals such as mice or fruit flies, the completion of the project has revealed that the human genome has an unexpectedly low gene density. Estimates of the number of genes in a genome are difficult to compile because they depend on gene finding algorithms that search for patterns resembling those present in known genes, such as open reading frames, promoter regions with sequences resembling the consensus promoter sequence, and related regulatory regions such as TATA boxes in eukaryotes. Gene finding is less reliable in eukaryotic than in prokaryotic genomes due to the presence of non-coding DNA such as introns and pseudogenes. Computational gene finding methods are still significantly more reliable than earlier techniques that required mapping the locations of specific mutations that gave rise to distinguishable alleles.

Gene Content and Genome Size of Various Organisms

Species	*Genome size (Mb)*	*Number of genes*
Mycoplasma genitalium	0.58	500
Streptococcus pneumoniae	2.2	2300

Escherichia coli	4.6	4400
Saccharomyces cerevisiae	12	5800
Arabidopsis thaliana	125	25,500
Caenorhabditis elegans	97	19,000
Sea urchin	814	23,300
Drosophila melanogaster	180	13,700
Mus musculus	2500	29,000
Homo sapiens	2900	27,000
Oryza sativa	466	45-55,000

In most eukaryotic species, very little of the DNA in the genome encodes proteins, and the genes may be separated by vast regions of non-coding DNA, much of which has been labeled "junk DNA" due to its apparent lack of function in the modern organism. A commonly studied type of "junk DNA" is the pseudogenes, or region of non-coding DNA that resembles expressed genes but usually lacks appropriate promoters and other control sequences; such regions are hypothesized to be the results of gene duplication events in a lineage's evolutionary past. Moreover, the genes are often fragmented internally by non-coding sequences called introns, which can be many times longer than the coding sequence but are spliced during post-transcriptional modification of pre-mRNA.

Genetic and Genomic Nomenclature

Gene nomenclature has been established by the HUGO Gene Nomenclature Committee (HGNC) for each known human gene in the form of an approved gene name and symbol (short-form abbreviation). All approved symbols are stored in the HGNC Database. Each symbol is unique and each gene is only given one approved gene symbol. It is necessary to provide a unique symbol for each gene so that people can talk about them. This also facilitates electronic data retrieval from publications. In preference each symbol maintains parallel construction in different members of a gene family and can be used in other species, especially the mouse.

Evolutionary Concept of a Gene

George C. Williams first explicitly advocated the gene-centric view of evolution in his 1966 book *Adaptation and Natural Selection*. He proposed an evolutionary concept of gene to be used when we are talking about natural selection favoring some genes. The definition is: "that which segregates and recombines with appreciable frequency." According to this definition, even an asexual genome could be considered a gene, insofar it have an appreciable permanency through many generations.

The difference is: the molecular gene *transcribes* as a unit, and the evolutionary gene *inherits* as a unit.

Richard Dawkins' *The Selfish Gene* and *The Extended Phenotype* defended the idea that the gene is the only replicator in living systems. This means that only genes transmit their structure largely intact and are potentially immortal in the form of copies. So, genes should be the unit of selection. In *The Selfish Gene* Dawkins attempts to redefine the word 'gene' to mean "an inheritable unit" instead of the generally accepted definition of "a section of DNA coding for a particular protein". In *River Out of Eden*, Dawkins further refined the idea of gene-centric selection by describing life as a river of compatible genes flowing through geological time. Scoop up a bucket of genes from the river of genes, and we have an organism serving as temporary bodies or survival machines. A river of genes may fork into two branches representing two non-interbreeding species as a result of geographical separation.

The *Gene* Concept is Still Changing

The concept of the gene has changed considerably. Originally considered a "unit of inheritance" to a usually DNA-based unit that can exert its effects on the organism through RNA or protein products. It was also previously believed that one gene makes one protein; this concept has been overthrown by the discovery of alternative splicing and trans-splicing.

And the definition of gene is still changing. The first cases of RNA-based inheritance have been discovered in mammals. In plants, cases of traits reappearing after several generations of absence have lead researchers to hypothesise RNA-directed overwriting of genomic DNA. Evidence is also accumulating that the control regions of a gene do not necessarily have to be close to the coding sequence on the linear molecule or even on the same chromosome. Spilianakis and colleagues discovered that the promoter region of the interferon-gamma gene on chromosome 10 and the regulatory regions of the T(H)2 cytokine locus on chromosome 11 come into close proximity in the nucleus possibly to be jointly regulated.

The concept that genes are clearly delimited is also being eroded. There is evidence for fused proteins stemming from two adjacent genes that can produce two separate protein products. While it is not clear whether these fusion proteins are functional, the phenomena is more frequent than previously thought. Even more ground-breaking than the discovery of fused genes is the observation that some proteins can be composed of exons from far away regions and even different chromosomes. This new data has led to an updated, and probably tentative, definition of a gene as "a union of genomic sequences encoding a coherent set of potentially overlapping functional products." This new definition categorizes genes by functional products, whether they be proteins or RNA, rather than specific DNA loci; all regulatory elements of DNA are therefore classified as *gene-associated* regions.

BIOMOLECULE

A biomolecule is a molecule that naturally occurs in living organisms. Biomolecules consist primarily of carbon and hydrogen, along with nitrogen, oxygen, phosphorus and sulfur. Other elements sometimes are incorporated but are much less common.

Explanation

All known forms of life are composed solely of biomolecules. For example, humans possess skin and hair. The main component of hair is keratin, an agglomeration of proteins which are themselves polymers built from amino acids. Amino acids are some of the most important building blocks used in nature to construct larger molecules. Another type of building block are the nucleotides, each of which consists of three components: either a purine or pyrimidine base, a pentose sugar and a phosphate group. These nucleotides mainly form the nucleic acids.

Besides the polymeric biomolecules, numerous organic molecules are absorbed by living systems.

Types of Biomolecules

A diverse range of biomolecules exist, including:

- Small molecules:
- Lipid, Phospholipid, Glycolipid, Sterol
- Vitamin
- Hormone, Neurotransmitter
- Carbohydrate, Sugar
- Disaccharide
- Monomers:
- Amino acid
- Nucleotide
- Phosphate
- Monosaccharide
- Polymers:

- Peptide, Oligopeptide, Polypeptide, Protein
- Nucleic acid, i.e. DNA, RNA
- Oligosaccharide, Polysaccharide

Nucleosides and Nucleotides

Nucleosides are molecules formed by attaching a nucleobase to a ribose ring. Examples of these include cytidine, uridine, adenosine, guanosine, thymidine and inosine.

Nucleosides can be phosphorylated by specific kinases in the cell, producing nucleotides, which are the molecular building blocks of DNA (deoxyribonucleic acid) and RNA (ribonucleic acid).

Saccharides

Monosaccharides are carbohydrates in the form of simple sugars. Examples of monosaccharides are the hexoses glucose, fructose, and galactose and pentoses, ribose, and deoxyribose

Disaccharides are formed from two monosaccharides joined together. Examples of disaccharides include sucrose, maltose, and lactose

Monosaccharides and disaccharides are sweet, water soluble, and crystalline.

Polysaccharides are polymerized monosaccharides, complex, unsweet carbohydrates. Examples are starch, cellulose, and glycogen. They are generally large and often have a complex, branched, connectivity. They are insoluble in water and do not form crystals. Shorter polysaccharides, with 2-15 monomers, are sometimes known as oligosaccharides.

Lipids

Lipids are chiefly fatty acid esters, and are the basic building blocks of biological membranes. Another biological role is energy storage (e.g., triglycerides). Most lipids consist of a polar or hydrophilic head (typically glycerol) and one to three nonpolar or hydrophobic fatty acid tails, and therefore they are amphiphilic. Fatty acids consist of unbranched chains of carbon atoms that are connected by single bonds alone (saturated fatty acids) or by both single and double bonds (unsaturated fatty acids). The chains are usually 14-24 carbon groups long, but it is always an even number.

For lipids present in biological membranes, the hydrophilic head is from one of three classes:

- Glycolipids, whose heads contain an oligosaccharide with 1-15 saccharide residues.
- Phospholipids, whose heads contain a positively charged group that is linked to the tail by a negatively charged phosphate group.
- Sterols, whose heads contain a planar steroid ring, for example, cholesterol.

Other lipids include prostaglandins and leukotrienes which are both 20-carbon fatty acyl units synthesized from arachidonic acid. They are also known as fatty acids

Hormones

Hormones are produced in the endocrine glands, where they are secreted into the bloodstream. They perform a wide range of roles in the various organs including the regulation of metabolic pathways and the regulation of membrane transport processes.

Hormones may be grouped into three structural classes:

- The steroids are one class of such hormones. They perform a variety of functions, but they are all made from cholesterol.
- Simple amines or amino acids.
- Peptides or proteins.

Amino Acids

Amino acids are molecules that contain both amino and carboxylic acid functional groups. (In biochemistry, the term amino acid is used when referring to those amino acids in which the amino and carboxylate functionalities are attached to the same carbon, plus proline which is not actually an amino acid).

Amino acids are the building blocks of long polymer chains. With 2-10 amino acids such chains are called peptides, with 10-100 they are often called polypeptides, and longer chains are known as proteins. These protein structures have many structural and functional roles in organisms.

There are twenty amino acids that are encoded by the standard genetic code, but there are more than 500 natural amino acids. When amino acids other than the set of twenty are observed in proteins, this is usually the result of modification after translation (protein synthesis). Only two amino acids other than the standard twenty are known to be incorporated into proteins during translation, in certain organisms:

- Selenocysteine is incorporated into some proteins at a UGA codon, which is normally a stop codon.

- Pyrrolysine is incorporated into some proteins at a UAG codon. For instance, in some methanogens in enzymes that are used to produce methane.

Besides those used in protein synthesis, other biologically important amino acids include carnitine (used in lipid transport within a cell), ornithine, GABA and taurine.

Protein Structure

The particular series of amino acids that form a protein is known as that protein's primary structure. Proteins have several, well-classified, **elements of local** structure and these are termed secondary structure. The overall 3D **structure of a** protein is termed its tertiary structure. Proteins often aggregate into **macromolecular** structures, or quaternary structure.

Metalloproteins

A metalloprotein is a molecule that contains a metal cofactor. The **metal attached** to the protein may be an isolated ion or may be a complex organometallic compound or organic compound, such as the porphyrin group found in hemoproteins. In some cases, the metal is coordinated with both a side chain of the protein and an inorganic nonmetallic ion. This type of protein-metal-nonmetal structure is found in iron-sulfur clusters.

Vitamins

A vitamin is a compound that cannot be synthesized by a given organism but is nonetheless vital to its survival or health (for example coenzymes). These compounds must be absorbed, or eaten, but typically only in trace quantities. When originally discovered by a Polish doctor, he believed them to all be basic. He therefore named them vital amines. The l was dropped to form the word vitamines.

MOLECULAR BIOLOGY

Molecular biology is the study of biology at a molecular level. The field overlaps with other areas of biology and chemistry, particularly genetics and biochemistry. Molecular biology chiefly concerns itself with understanding the interactions between the various systems of a cell, including the interrelationship of DNA, RNA and protein biosynthesis and learning how these interactions are regulated.

Writing in *Nature*, William Astbury described molecular biology as:

> "... not so much a technique as an approach, an approach from the viewpoint of the so-called basic sciences with the leading idea of searching below the large-scale manifestations of classical biology for the corresponding molecular plan. It

is concerned particularly with the forms of biological molecules and ..... is predominantly three-dimensional and structural - which does not mean, however, that it is merely a refinement of morphology - it must at the same time inquire into genesis and function."

Relationship to Other "Molecular-Scale" Biological Sciences

Researchers in molecular biology use specific techniques native to molecular biology (see *Techniques* section later in article), but increasingly combine these with techniques and ideas from genetics and biochemistry. There is not a defined line between these disciplines. Today the terms molecular biology and biochemistry are nearly interchangeable. The following figure is a schematic that depicts one possible view of the relationship between the fields:

- *Biochemistry* is the study of the chemical substances and vital processes occurring in living organisms. Biochemists focus heavily on the role, function, and structure of biomolecules. The study of the chemistry behind biological processes and the synthesis of biologically active molecules are examples of biochemistry.
- *Genetics* is the study of the effect of genetic differences on organisms. Often this can be inferred by the absence of a normal component (e.g. one gene). The study of "mutants" – organisms which lack one or more functional components with respect to the so-called "wild type" or normal phenotype. Genetic interactions such as epistasis can often confound simple interpretations of such "knock-out" studies.
- *Molecular biology* is the study of molecular underpinnings of the process of replication, transcription and translation of the genetic material. The central dogma of molecular biology where genetic material is transcribed into RNA and then translated into protein, despite being an oversimplified picture of molecular biology, still provides a good starting point for understanding the field. This picture, however, is undergoing revision in light of emerging novel roles for RNA.

Much of the work in molecular biology is quantitative, and recently much work has been done at the interface of molecular biology and computer science in bioinformatics and computational biology. As of the early 2000s, the study of gene structure and function, molecular genetics, has been amongst the most prominent sub-field of molecular biology.

Increasingly many other fields of biology focus on molecules, either directly studying their interactions in their own right such as in cell biology and developmental biology, or indirectly, where the techniques of molecular biology are used to infer historical attributes of populations or species, as in fields in evolutionary biology such as population genetics and phylogenetics. There is also a long tradition of studying biomolecules "from the ground up" in biophysics.

Techniques of Molecular Biology

Since the late 1950s and early 1960s, molecular biologists have learned to characterize, isolate, and manipulate the molecular components of cells and organisms. These components include DNA, the repository of genetic information; RNA, a close relative of DNA whose functions range from serving as a temporary working copy of DNA to actual structural and enzymatic functions as well as a functional and structural part of the translational apparatus; and proteins, the major structural and enzymatic type of molecule in cells.

Expression Cloning

One of the most basic techniques of molecular biology to study protein function is expression cloning. In this technique, DNA coding for a protein of interest is cloned (using PCR and/or restriction enzymes) into a plasmid (known as an expression vector). This plasmid may have special promoter elements to drive production of the protein of interest, and may also have antibiotic resistance markers to help follow the plasmid.

This plasmid can be inserted into either bacterial or animal cells. Introducing DNA into bacterial cells is called transformation, and can be completed with several methods, including electroporation, microinjection, passive uptake and conjugation. Introducing DNA into eukaryotic cells, such as animal cells, is called transfection. Several different transfection techniques are available, including calcium phosphate transfection, liposome transfection, and proprietary transfection reagents such as Fugene. DNA can also be introduced into cells using viruses or pathenogenic bacteria as carriers. In such cases, the technique is called viral/bacterial transduction, and the cells are said to be transduced.

In either case, DNA coding for a protein of interest is now inside a cell, and the protein can now be expressed. A variety of systems, such as inducible promoters and specific cell-signaling factors, are available to help express the protein of interest at high levels. Large quantities of a protein can then be extracted from the bacterial or eukaryotic cell. The protein can be tested for enzymatic activity under a variety of situations, the protein may be crystallized so its tertiary structure can be studied, or, in the pharmaceutical industry, the activity of new drugs against the protein can be studied.

Polymerase Chain Reaction (PCR)

The polymerase chain reaction is an extremely versatile technique for copying

DNA. In brief, PCR allows a single DNA sequence to be copied (millions of times), or altered in predetermined ways. For example, PCR can be used to introduce restriction enzyme sites, or to mutate (change) particular bases of DNA. PCR can also be used to determine whether a particular DNA fragment is found in a cDNA library. PCR has many variations, like reverse transcription PCR (RT-PCR) for amplification of RNA, and, more recently, real-time PCR (qPCR) which allow for quantitative measurement of DNA or RNA molecules.

Gel Electrophoresis

Gel electrophoresis is one of the principal tools of molecular biology. The basic principle is that DNA, RNA, and proteins can all be separated using an electric field. In agarose gel electrophoresis, DNA and RNA can be separated based on size by running the DNA through an agarose gel. Proteins can be separated based on size using an SDS-PAGE gel, or by size and their electric charge, using what is known as a 2d gel.

Southern Blotting

Named after its inventor, biologist Edwin Southern, the Southern blot is a method for probing for the presence of a specific DNA sequence within a DNA sample. DNA samples before or after restriction enzyme digestion are separated by gel electrophoresis and then transferred to a membrane by blotting via capillary action. The membrane can then be probed using a DNA probe labeled using a complement of the sequence of interest. Most original protocols used radioactive labels, however now non-radioactive alternatives are available. Southern blotting is less commonly used in laboratory science due to the capacity of using PCR to detect specific DNA sequences from DNA samples. However, these blots are still used for some applications, such as measuring transgene copy number in transgenic mice, or in the engineering of gene knockout embryonic stem cell lines.

Northern Blotting

The Northern blot is used to study the expression patterns a specific type of RNA molecule as relative comparison among of a set of different samples of RNA. It is essentially a combination of denaturing RNA gel electrophoresis, and a blot. In this process RNA is separated based on size and is then transferred to a membrane that is then probed with a labeled complement of a sequence of interest. The results may be visualized through a variety of ways depending on the label used, however, most result in the revelation of bands representing the sizes of the RNA detected in sample. The intensity of these bands is related to the amount of the target RNA in the samples analyzed. The procedure is commonly used to study when and how much gene expression is occurring by measuring how much of that RNA is present in different samples. It is one of the most basic tools for determining at what time, and under what conditions, certain genes are expressed in living tissues.

Western Blotting

Antibodies to most proteins can be created by injecting small amounts of the protein into an animal such as a mouse, rabbit, sheep, or donkey (polyclonal antibodies)or produced in cell culture (monoclonal antibodies). These antibodies can be used for a variety of analytical and preparative techniques.

In western blotting, proteins are first separated by size, in a thin gel sandwiched between two glass plates in a technique known as SDS-PAGE (sodium dodecyl sulphate polyacrylamide gel electrophoresis). The proteins in the gel are then transferred to a PVDF, nitrocellulose, nylon or other support membrane. This membrane can then be probed with solutions of antibodies. Antibodies that specifically bind to the protein of interest can then be visualized by a variety of techniques, including coloured products, chemiluminescence, or autoradiography.

Analogous methods to western blotting can also be used to directly stain specific proteins in cells and tissue sections. However, these *immunostaining* methods are typically more associated with cell biology than molecular biology.

The terms “western” and “northern” are jokes: The first blots were with DNA, and since they were done by Ed Southern, they came to be known as Southerns. Patricia Thomas, inventor of the RNA blot, which became known as a “northern”, actually didn’t use the term. To carry the joke further, one can find reference in the literature to “southwesterns” (Protein-DNA interactions) and “farwesterns” (Protein-Protein interactions).

Arrays

A DNA array is a collection of spots attached to a solid support such as a microscope slide; each spot contains one or more DNA oligonucleotides. Arrays make it possible to put down a large number of very small (100 micrometre diameter) spots on a single slide; if each spot has a DNA molecule that is complementary to a single gene (similar to Southern blotting), one can analyze the expression of every gene in an organism in a single experiment. For instance, the common baker’s yeast, *Saccharomyces cerevisiae*, contains about 7000 genes; with a microarray, one can measure quantitatively, how each gene is expressed, and how that expression changes, for example, with a change in temperature. There are many different ways to fabricate microarrays; the most common are silicon chips, microscope slides with spots of ~ 100 micrometre diameter, custom arrays, and arrays with larger spots on porous membranes (macroarrays).

Arrays can also be made with molecules other than DNA. For example, an antibody array can be used to determine what proteins or bacteria are present in a blood sample.

Abandoned Technology

As new procedures and technology become available, the older technology is rapidly abandoned. A good example is methods for determining the size of DNA molecules. Prior to gel electrophoresis (agarose or polyacrylamide) DNA was sized with rate sedimentation in sucrose gradients, a slow and labor intensive technology requiring expensive instrumentation; prior to sucrose gradients, viscometry was used.

Aside from their historical interest, it is worth knowing about older technology as it may be useful to solve a particular problem.

History

Molecular biology was established in the 1930s, the term was first coined by Warren Weaver in 1938 however. Warren was director of Natural Sciences for the Rockefeller Foundation at the time and believed that biology was about to undergo a period of significant change given recent advances in fields such as X-ray crystallography. He therefore channeled significant amounts of (Rockefeller Institute) money into biological fields.

HOMEOSTASIS

Homeostasis is the property of either an open system or a closed system, especially a living organism, which regulates its internal environment so as to maintain a stable, constant condition. Multiple dynamic equilibrium adjustments and regulation mechanisms make homeostasis possible. The concept was created by Claude Bernard, often considered as the father of physiology, and published in 1865. The term was coined in 1932 by Walter Bradford Cannon from the Greek *homoios* (same, like, resembling) and *stasis* (to stand, posture).

Biological Homeostasis

With regard to any given life system parameter, an organism may be a *conformer* or a *regulator*. Regulators try to maintain the parameter at a constant level over possibly wide ambient environmental variations. On the other hand, conformers allow the environment to determine the parameter. For instance, endothermic animals maintain a constant body temperature, while ectothermic animals exhibit wide body temperature variation. Examples of endothermic animals include mammals and birds, examples of ectothermic animals include reptiles and some sea creatures.

This is not to say that conformers don't have behavioural adaptations allowing them to exert some control over a given parameter. For instance, reptiles often rest on sun-heated rocks in the morning to raise their body temperature. Likewise, regulators' behaviors may contribute to their internal stability: The same sun-baked rock may host a ground squirrel, also basking in the morning sun.

An advantage of homeostatic regulation is that it allows an organism to function effectively in a broad range of environmental conditions. For example, ectotherms tend to become sluggish at low temperatures, while a co-located endotherm may be fully active. That thermal stability comes at a price since an automatic regulation system requires additional energy. One reason snakes may eat only once a week is that they use much less energy to maintain homeostasis.

Most homeostatic regulation is controlled by the release of hormones into the bloodstream. However other regulatory processes rely on simple diffusion to maintain a balance.

Homeostatic regulation extends far beyond the control of temperature. All animals also regulate their blood glucose, as well as the concentration of their blood. Mammals regulate their blood glucose with insulin and glucagon. These hormones are released by the pancreas. If the pancreas is for any reason unable to produce enough of these two hormones diabetes results. The kidneys are used to remove excess water and ions from the blood. These are then expelled as urine. The kidneys perform a vital role in homeostatic regulation in mammals removing excess water, salt and urea from the blood. These are the body's main waste products.

Sleep timing depends upon a balance between homeostatic sleep propensity, the need for sleep as a function of the amount of time elapsed since the last adequate sleep episode, and circadian rhythms which determine the ideal timing of a correctly structured and restorative sleep episode.

Control Mechanisms

All homeostatic control mechanisms have at least three interdependent components for the variable being regulated: The receptor is the sensing component that monitors and responds to changes in the environment. When the receptor senses a stimulus, it sends information to a control center, the component that sets the range at which a variable is maintained. The control center determines an appropriate response to the stimulus. The result of that response feeds to the receptor, either enhancing it with positive feedback or depressing it with negative feedback

Negative Feedback Mechanisms

Negative feedback mechanisms reduce or suppress the original stimulus, given the effector's output. Most homeostatic control mechanisms require a negative feedback loop to keep conditions from exceeding tolerable limits. The purpose is to prevent sudden severe changes within a complex organism. There are hundreds of negative feedback mechanisms in the human body. Among the most important regulatory functions are: thermoregulation, osmoregulation, and glucoregulation. The kidneys contribute to homeostasis in five important ways: regulation of blood water levels, reabsorption of substances into the blood, maintenance of salt and ion levels in the blood, regulation of blood pH, and excretion of urea and other wastes.

A negative feedback mechanism example is the typical home heating system. Its thermostat houses a thermometer, the receptor that senses when the temperature is too low. The control center, also housed in the thermostat, senses and responds to the thermometer when the temperature drops below a specified set point. Below that target level, the thermostat sends a message to the effector, the furnace. The furnace then produces heat, which warms the house. Once the thermostat senses a target level of heat has been reached, it will signal the furnace to turn off, thus maintaining a comfortable temperature - not too hot nor cold.

Positive Feedback Mechanisms

Positive feedback mechanisms are designed to accelerate or enhance the output created by a stimulus that has already been activated.

Unlike negative feedback mechanisms that initiate to maintain or regulate physiological functions within a set and narrow range, the positive feedback mechanisms are designed to push levels out of normal ranges. To achieve this purpose, a series of events initiates a cascading process that builds to increase the effect of the stimulus. This process can be beneficial but is rarely used by the body due to risks of the acceleration becoming uncontrollable.

One bodily positive feedback example event is blood platelet accumulation which in turn causes blood clotting in response to a break or tear in the lining of blood vessels. Another example is the release of oxytocin to intensify the contractions that take place during childbirth.

Positive feedback can also be harmful. An example being when you have a fever it causes a positive feedback within homeostasis that pushes the temperature continually higher. Body temperature can reach extremes of 45°C (113°F), at which cellular proteins denature, causing the active site in proteins to change, thus causing metabolism stop and ultimately resulting in death.

Homeostatic Imbalance

Much disease results from disturbance of homeostasis, a condition known as homeostatic imbalance. As it ages, every organism will lose efficiency in its control systems. The inefficiencies gradually result in an unstable internal environment that increases the risk for illness. In addition, homeostatic imbalance is also responsible for the physical changes associated with aging. Even more serious than illness and other characteristics of ageing, is death. Heart failure has been seen where nominal negative feedback mechanisms become overwhelmed, and destructive positive feedback mechanisms then take over.

Diseases which result from a homeostatic imbalance include diabetes, dehydration, hypoglycemia, hyperglycemia, gout and any disease caused by a toxin present in the bloodstream. All of these conditions result from the presence of an increased amount of a particular substance. Ideally homeostatic control mechanisms should prevent this imbalance from occurring but in some people the mechanisms do not work efficiently enough or the quantity of the substance exceeds the levels at which it can be managed. In these cases medical intervention is necessary to restore the imbalance or permanent damage to the organs may result.

Varieties of Homeostasis

The Dynamic Energy Budget theory for metabolic organisation delineates structure and (one or more) reserves in an organism. Its formulation is based on three forms of homeostasis:

- Strong homeostasis is where structure and reserve do not change in composition. Since the amount of reserve and structure can vary, this allows a particular change in the composition of the whole body (as explained by the Dynamic Energy Budget theory).
- Weak homeostasis is where the ratio of the amounts of reserve and structure becomes constant as long as food availability is constant, even when the organism grows. This means that the whole body composition is constant during growth in constant environments.
- Structural homeostasis means that the sub-individual structures grow in harmony with the whole individual; the relative proportions of the individuals remain constant.

Ecological Homeostasis

Ecological homeostasis is found in a climax community of maximum permitted biodiversity, given the prevailing ecological conditions.

In disturbed ecosystems or sub-climax biological communities such as the island of Krakatoa, after its major eruption in 1883, the established stable homeostasis of the previous forest climax ecosystem was destroyed and all life eliminated from the island. In the years after the eruption, Krakatoa went through a sequence of ecological changes in which successive groups of new plant or animal species followed one another, leading to increasing biodiversity and eventually culminating in a re-established climax community. This *ecological succession* on Krakatoa occurred in a number of several stages, in which a *sere* is defined as "a stage in a sequence of events by which succession occurs". The complete chain of seres leading to a climax is called a *prisere*. In the case of Krakatoa, the island as reached its climax community with eight hundred different species being recorded in 1983, one hundred years after the eruption which cleared all life off the island. Evidence confirms that this number has been homeostatic for some time, with the introduction of new species rapidly leading to elimination of old ones.

The evidence of Krakatoa, and other disturbed or virgin ecosystems shows that the initial colonisation by *pioneer* or *R strategy* species occurs through positive feedback reproduction strategies, where species are weeds, producing huge numbers of possible offspring, but investing little in the success of any one. Rapid boom and bust plague or pest cycles are observed with such species. As an ecosystem starts to approach climax these species get replaced by more sophisticated climax species which through negative feedback, adapt themselves to specific environmental conditions. These species, closely controlled by *carrying capacity*, follow *K strategies* where species produce fewer numbers of potential offspring, but invest more heavily in securing the reproductive success of each one to the micro-environmental conditions of its specific ecological niche.

It begins with a pioneer community and ends with a climax community. This climax community occurs when the ultimate vegetation has become in equilibrium with the local environment.

Such ecosystems form nested communities or *heterarchies*, in which homeostasis at one level, contributes to homeostatic processes at another holonic level. For example, the loss of leaves on a mature rainforest tree gives a space for new growth, and contributes to the plant litter and soil humus build-up upon which such growth depends. Equally a mature rainforest tree reduces the sunlight falling on the forest floor and helps prevent invasion by other species. But trees too fall to the forest floor and a healthy forest glade is dependent upon a constant rate of forest regrowth, produced by the fall of logs, and the recycling of forest nutrients through the respiration of termites and other insect, fungal and bacterial decomposers. Similarly such forest glades contribute ecological services, such as the regulation of microclimates or of the hydrological cycle for an ecosystem, and a number of different ecosystems act together to maintain homeostasis perhaps of a number of river catchments within a bioregion. A diversity of bioregions similarly makes up a stable homeostatic biological region or biome.

In the Gaia hypothesis, James Lovelock stated that the entire mass of living matter on Earth (or any planet with life) functions as a vast homeostatic superorganism that actively modifies its planetary environment to produce the environmental conditions necessary for its own survival. In this view, the entire planet maintains homeostasis. Whether this sort of system is present on Earth is still open to debate. However, some relatively simple homeostatic mechanisms are generally accepted. For example, when atmospheric carbon dioxide levels rise, certain plants are able to grow better and thus act to remove more carbon dioxide from the atmosphere. When sunlight is plentiful and atmospheric temperature climbs, the phytoplankton of the ocean surface waters thrive and produce more dimethyl sulfide, DMS. The DMS molecules act as cloud condensation nuclei which produce more clouds and thus increase the atmospheric albedo and this feeds back to lower the temperature of the atmosphere. As scientists discover more about Gaia, vast numbers of positive and negative feedback loops are being discovered, that together maintain a metastable condition, sometimes within very broad range of environmental conditions.

Reactive Homeostasis

Example of use: "Reactive homeostasis is an immediate response to a homeostatic challenge such as predation."

However, any homeostasis is impossible without reaction - because homeostasis is and must be a "feedback" phenomenon.

The phrase "reactive homeostasis" is simply short for: "reactive compensation reestablishing homeostasis", that is to say, "reestablishing a point of homeostasis." - it should not be confused with a separate *kind* of homeostasis or a distinct phenomenon *from* homeostasis, it is simply the compensation (or compensatory) phase of homeostasis.

Other Fields

The term has come to be used in other fields, as well.

Risk Homeostasis

An actuary may refer to *risk homeostasis*, where (for example) people who have anti-lock brakes have no better safety record than those without anti-lock brakes, because they unconsciously compensate for the safer vehicle via less-safe driving habits. Previously, certain maneuvers involved minor skids, evoking fear and avoidance:

now the anti-lock system moves the boundary for such feedback, and behavior patterns expand into the no-longer punitive area. It has also been suggested that ecological crises are an instance of risk homeostasis in which behavior known to be dangerous continues until dramatic consequences actually occur.

Stress Homeostasis

Sociologists and psychologists may refer to *stress homeostasis*, the tendency of a population or an individual to stay at a certain level of stress, often generating artificial stresses if the "natural" level of stress is not enough.

Jean Francois Lyotard, a postmodern theorist, has applied this term to societal 'power centers' that he describes as being 'overned by a principle of homeostasis.' For example the scientific hierarchy, which will sometimes ignore a radical new discovery for years because it destabilizes previously accepted norms.

Waste Homeostasis

Andrew Potter has used the term *waste homeostasis* in reference to the lack of net gain from energy saving technologies.

Conversational Homeostasis

A 2007 study purported to find (and show clinically) *conversational homeostasis* in which overly-familiar people (such as spouses) condense their speech so much that they are actually worse at communicating novel information than strangers are, while not being conscious of this problem.

Metabolic Homeostasis

Some herbal medicines, known as adaptogens, have been defined to function as non-toxic metabolic regulators that can enhance metabolic homeostasis during stress.

REFERENCES

Braig M, Schmitt C (2006). "Oncogene-induced senescence: putting the brakes on tumor development". *Cancer Res* 66 (6): 2881-4.

Cavalier-Smith T. (1985). Eukaryotic gene numbers, non-coding DNA, and genome size. In Cavalier-Smith T, ed. *The Evolution of Genome Size* Chichester: John Wiley.

Cohen, S.N., Chang, A.C.Y., Boyer, H. & Heling, R.B. Construction of biologically functional bacterial plasmids *in vitro*.

Darwin C. (1868). Animals and Plants under Domestication (1868).

Elizabeth Pennisi (2007). "DNA Study Forces Rethink of What It Means to Be a Gene". *Science* 316 (5831): 1556-1557.

Gerstein MB, Bruce C, Rozowsky JS, Zheng D, Du J, Korbel JO, Emanuelsson O, Zhang ZD, Weissman S, Snyder M (2007). "What is a gene, post-ENCODE? History and updated definition". *Genome Research* 17 (6): 669-681.

International Human Genome Sequencing Consortium (2004). "Finishing the euchromatic sequence of the human genome.". *Nature* 431 (7011): 931-45.

Kapranov & colleagues (2005) Examples of the complex architecture of the human transcriptome revealed by RACE and high-density tiling arrays.

Keysar, Boaz (2007), "The Effect of Information Overlap on Communication Effectiveness," Cognitive Science.

Lodish, H, Berk A, Matsudaira P, Kaiser CA, Krieger M, Scott MP, Zipursky SL, Darnell J. (2004). *Molecular Cell Biology*, 5th, New York: WH Freeman.

Lolle & colleagues (2005) Genome-wide non-mendelian inheritance of extra-genomic information in Arabidopsis.

Marieb, Elaine N. and Hoehn, Katja (2007). *Human Anatomy an Physiology* (Seventh ed.). San Francisco, CA: Pearson Benjamin Cummings.

Mark B. Gerstein *et al.*, "What is a gene, post-ENCODE? History and updated definition," *Genome Research* 17(6) (2007): 669-681.

Min Jou W, Haegeman G, Ysebaert M, Fiers W (1972). "Nucleotide sequence of the gene coding for the bacteriophage MS2 coat protein". *Nature* 237 (5350): 82-8.

Mount, DW (2004). *Bioinformatics: Sequence and genome analysis*, 2nd ed., Cold Spring Harbor Laboratory Press: Cold Spring Harbor, New York.

Parra & colleagues (2006) Tandem chimerism as a means to increase protein complexity in the human genome.

Pearson H (2006). "Genetics: what is a gene?". *Nature* 441 (7092): 398-401.

Pennisi, Elizabeth (2007). "Working the (Gene Count) Numbers_ Finally, a Firm Answer". *Science* 316 (5828): 1113.

Potter, Andrew (2007), "Planet-Friendly design? Bah, humbug.", MacLean's 120 (5): 14.

Rassoulzadegan M, Grandjean V, Gounon P, Vincent S, Gillot I, Cuzin F (2006). "RNA-mediated non-mendelian inheritance of an epigenetic change in the mouse". *Nature* 441 (7092): 469-74.

Rodgers, M. The Pandora's box congress. *Rolling Stone* 189, 37 – 77 (1975).

Spilianakis & colleagues (2005) Interchromosomal associations between alternatively expressed loci.

The Human Genome Project Timeline. Retrieved on 2006-09-13.

Vries, H. de (1889) *Intracellular Pangenesis* ("pangen" definition on page 7 and 40 of this 1910 translation in English).

Watson JD, Baker TA, Bell SP, Gann A, Levine M, Losick R (2004). *Molecular Biology of the Gene*, 5th ed., Peason Benjamin Cummings (Cold Spring Harbor Laboratory Press).

Winston, David and Maimes, Steven. "Adaptogens: Herbs for Strength, Stamina, and Stress Relief," Healing Arts Press, 2007.

Woodson SA (1998). "Ironing out the kinks: splicing and translation in bacteria". *Genes Dev.* 12 (9): 1243–7.

Wyatt, James K.; Ritz-De Cecco, Angela; Czeisler, Charles A.; Dijk, Derk-Jan (October 1999). "Circadian temperature and melatonin rhythms, sleep, and neurobehavoral function in hman living on a 20-h day." *Am J Physiol* 277 (4): R1152-R1163. Fulltext. Retrieved on 2007-11-25. "... significant homeostatic and circadian modulation of sleep structure, with the highest sleep efficiency occurring in sleep episodes bracketing the melatonin maximum and core body temperature minimum."

2

Modern Bioscience Branches and Areas of Studies

MODERN BIOSCIENCE

Modern bioscience and medical technology began in the mid-1970s. Using biology and the other life sciences, bioscience and medical technology produces and improves health care, agriculture and environmental products by working with deoxyribonucleic acid (DNA) the basic building blocks of all living organisms and antibodies, natural disease fighters. Modern bioscience and medical technology can be broken into several distinct fields. Bioscience began nearly 10,000 years ago when agriculture replaced hunting and gathering. People began producing wine, beer and bread by using and manipulating natural processes. Herds of livestock were improved with primitive animal husbandry techniques.

Bioinformatics

Bioinformatics is the use of information technology to store and analyze genetic information. Bioinformatic researchers develop and apply computing tools to extract the secrets of the life and death of organisms from the genetic blueprints and molecular structure stored in digital collections. Traditionally, biology research begins with a hypothesis. A biologist then collects experimental data and analyzes them to support or disprove the hypothesis. However, information technology is changing this sequence of events. Today, large-scale exploratory experiments are gathering as much data as possible. The Human Genome Project, for example, is creating an inventory of all 3 billion amino acids in the human genetic blueprint. So now when a biologist forms a hypothesis, the data may already be in such a collection, just a computer search away. Bioinformatics is expected to help scientists discover the genetic basis of many diseases and accelerate the development of more effective pharmaceutics to combat them.

Bio-agriculture

Bio-agriculture improves the quality of seed grains to increase crop yields, increases protein levels in forage crops and creates species of plants that are more resistant to disease, insects and viruses, as well as drought, floods, temperature extremes and salinity. Bio-agriculture can help produce more nutritious foods containing higher levels of vitamin C, vitamin E, beta carotene and other substances the body needs. Bio-agriculture reduces production costs and helps the environment by reducing the use of pesticides, and reduces environmental degradation by reducing the need for irrigation.

Molecular and Cellular Biology

Basic research on cells which integrates biology, chemistry, engineering and computer science. There are two major areas of research. The first focuses on manipulating DNA to produce a desired effect. Example: Human insulin needed by diabetics can be made in a lab by cutting from a strand of human DNA the sequence that controls the production of insulin and pasting it to an E. Coli cell. The E. Coli cell will then produce insulin.

The second area of research focuses on producing customized disease- and infection-fighting antibodies similar to those the human body produces naturally to fight off common colds and other illnesses. The customized antibodies, called monoclonal antibodies, are designed to fight diseases our bodies cannot fight adequately. Example: a skin cancer vaccine.

Environmental Bioscience

Bioremediation uses microscopic organisms, whether naturally-occurring or genetically-altered, for a wide variety of applications in pollution control and prevention. Organisms can "eat" or, more accurately, degrade or transform toxic waste and oil spills into harmless byproducts. In some cases, nutrients are used to stimulate the growth of naturally-occurring organisms, while, in other cases, organisms are introduced to stimulate the process.

Environmental diagnostics use a combination of organisms and computer technology to create devices which can detect pollutants in water or air before they are released into the environment. This technology is being used by sewage treatment plants and coal-fired generators.

Gene Therapy

Gene therapy is the use of genes and the techniques of genetic engineering in the treatment of a genetic disorders or chronic disease. There are many techniques of gene therapy, most of them still in experimental stages.

One technique for treating hereditary problems involves removing cells from a patient, fortifying them with healthy copies of the defective gene, and reinjecting the fortified cells into the patient. Another involves inserting a gene into an inactivated or non-virulent virus and using the virus's infective capabilities to carry the desired gene into the patient's cells.

A liposome, a tiny fat-encased pouch that can traverse cell membranes, is also sometimes used to transport a gene into a cell.

Once inserted into body cells, the gene may produce an essential chemical that the patient's body cannot, remove or render harmless a substance or gene that is causing or contributing to disease, or expose certain cells, especially cancerous cells, to attack by conventional drugs. Like drugs, gene therapy techniques must be approved by the federal government.

There are four broad unifying principles of biology:

Homeostasis

Homeostasis is the property of an open system, especially living organisms, to regulate its internal environment to maintain a stable, constant condition, by means of multiple dynamic equilibrium adjustments, controlled by interrelated regulation mechanisms. The term was coined in 1932 by Walter Bradford Cannon from the Greek *homoios* (same, like, resembling) and *stasis* (to stand, posture).

Homeostasis is one of the fundamental characteristics of living things. It is the maintenance of the internal environment within tolerable limits. Factors such as temperature, salinity, acidity, and concentrations of nutrients and wastes, all affect the body's ability to sustain life.

With regard to any parameter, an organism may be a *conformer* or a *regulator*. Regulators try to maintain the parameter at a constant level, regardless of what is happening in its environment. Conformers allow the environment to determine the parameter. For instance, endothermic animals maintain a constant body temperature, while ectothermic animals exhibit wide variation in body temperature. This is not to say that conformers may not have behavioral adaptations that allow them to exert some control over the parameter in question. For instance, reptiles often sit on sun-heated rocks in the morning to raise their body temperatures.

An advantage of homeostatic regulation is that it allows the organism to function more effectively. For instance, ectotherms tend to become sluggish at low temperatures, whereas endotherms are as active as always. On the other hand, regulation requires energy. One reason snakes are able to eat just once a week is that they use much less energy for maintaining homeostasis.

All homeostatic control mechanisms have at least three interdependent components to the variable being regulated. The receptor, is the sensing component that monitors and responds to changes in the environment. Once the receptor senses a stimuli, it will send information to the control center, the component that sets the range at which a variable is maintained. The control center will determine an appropriate action based on the information it receives. The third component, the effector, will provide the output from the control center to the stimulus. The results of the response will feed back to the receptor, either enhancing it with positive feedback or depressing it with negative feed back(Marieb, 2007).

Negative feedback mechanisms stop the original stimulus, or at least reduce it, with the effector's output. Most of the homeostatic control mechanisms require negative feedback mechanisms to keep conditions from exceeding tolerable limits. The goal is to prevent sudden severe changes within the body. There are hundreds of negative feedback mechanisms in the human body, including: body temperature, blood volume, heart rate, blood levels of oxygen, carbon dioxide, and minerals.

A common example of how negative feedback mechanisms work is to imagine how a home heating system works. The thermostat houses a thermometer, the receptor that senses when the temperature is too low. The control center, which is also housed in the thermostat will sense and respond to the thermometer when the temperature drops below a specified level. When the temperature drops below the target level, the thermostat will send a message to the effector, the furnace. The furnace will then begin producing heat, which will heat the house. Once the thermostat senses the target level of heat, it will signal the furnace to turn off, keeping the house from becoming too warm. This example shows how negative feedback mechanisms keep the temperature in the house from becoming too cold or too warm (Marieb, 2007).

Positive Feedback Mechanisms are designed to accelerate or enhance the output created by a stimulus that has already been activated. Unlike negative feedback mechanisms that are initiated to maitian or regulate physiological functions within a set and narrow range, the positive feedback mechinisms are designed to push levels out of their normal ranges.To achieve this goal, a series of events will be started will start a cascading affect that will help to increase the effect of the stimulus. This process is beneficial, but is rarely used by the body because of the risks of the accceleration to become uncontrolable.

Two well known events that involve this process are the accumalation of blood platelets that cause blood clotting in response to a break or tear in the lining of blood vessels, as well as the release of oxytocin to intensify the contractions that take place during childbirth. (Marieb,2007)

The Dynamic Energy Budget theory for metabolic organisation delineates structure and (one or more) reserves in an organism. Its formulation is based on three different types of homeostasis:

- Weak homeostasis is that the ratio of the amounts of reserve and structure becomes constant as long as food availability is constant, even when the organism grows. This means that the whole body composition is constant during growth in constant environments.
- Strong homeostasis is that structure and reserve do not change in composition. Since the amount of reserve and structure can vary, this allows a particular change in the composition of the whole body as explained by the Dynamic Energy Budget theory.
- Structural homeostasis means that the sub-individual structures grow in harmony with the whole individual; the relative proportions of the individuals remain constant.

Ecological homeostasis is found in a *climax community* of maximum permitted biodiversity, given the prevailing ecological conditions. In disturbed ecosystems or sub-climax biological communities such as the island of Krakatoa, after its major eruption in 1883, the established stable homeostasis of the previous forest climax ecosystem was destroyed and all life eliminated from the island. In the years after the eruption, Krakatoa went through a sequence of ecological changes in which successive groups of new plant or animal species followed one another, leading to increasing biodiversity and eventually culminating in a re-established climax community.

This *ecological succession* on Krakatoa occurred in a number of several stages, in which a *sere* is defined as "a stage in a sequence of events by which succession occurs". The complete chain of seres leading to a climax is called a *prisere*. In the case of Krakatoa, the island as reached its climax community with eight hundred different species being recorded in 1983, one hundred years after the eruption which cleared all life off the island. Evidence confirms that this number has been homeostatic for some time, with the introduction of new species rapidly leading to elimination of old ones. The evidence of Krakatoa, and other disturbed or virgin ecosystems shows that the initial colonisation by *pioneer* or *R strategy* species occurs through positive feedback reproduction strategies, where species are weeds, producing huge numbers of possible offspring, but investing little in the success of any one.

Rapid boom and bust plague or pest cycles are observed with such species. As an ecosystem starts to approach climax these species get replaced by more sophisticated climax species which through negative feedback, adapt themselves to specific environmental conditions. These species, closely controlled by *carrying capacity*, follow *K strategies* where species produce fewer numbers of potential offspring, but invest more heavily in securing the reproductive success of each one to the micro-environmental conditions of its specific ecological niche. It begins with a pioneer community and ends with a climax community. This climax community occurs when the ultimate vegetation has become in equilibrium with the local environment.

Such ecosystems form nested communities or *heterarchies*, in which homeostasis at one level, contributes to homeostatic processes at another holonic level. For example, the loss of leaves on a mature rainforest tree gives a space for new growth, and contributes to the plant litter and soil humus build-up upon which such growth depends. Equally a mature rainforest tree reduces the sunlight falling on the forest floor and helps prevent invasion by other species.

But trees too fall to the forest floor and a healthy forest glade is dependent upon a constant rate of forest regrowth, produced by the fall of logs, and the recycling of forest nutrients through the respiration of termites and other insect, fungal and bacterial decomposers. Similarly such forest glades contribute ecological services, such as the regulation of microclimates or of the hydrological cycle for an ecosystem, and a number of different ecosystems act together to maintain homeostasis perhaps of a number of river catchments within a bioregion. A diversity of bioregions similarly makes up a stable homeostatic biological region or biome. In the Gaia hypothesis, James Lovelock stated that the entire mass of living matter on Earth (or any planet with life) functions as a vast homeostatic superorganism that actively modifies its planetary environment to produce the environmental conditions necessary for its own survival. In this view, the entire planet maintains homeostasis. Whether this sort of system is present on Earth is still open to debate.

However, some relatively simple homeostatic mechanisms are generally accepted. For example, when atmospheric carbon dioxide levels rise, certain plants are able to grow better and thus act to remove more carbon dioxide from the atmosphere. When sunlight is plentiful and atmospheric temperature climbs, the phytoplankton of the ocean surface waters thrive and produce more dimethyl sulfide, DMS. The DMS molecules act as cloud condensation nuclei which produce more clouds and thus increase the atmospheric albedo and this feeds back to lower the temperature of the atmosphere. As scientists discover more about Gaia, vast numbers of positive and negative feedback loops are being discovered, that together maintain a metastable condition, sometimes within very broad range of environmental conditions. Example of use: "Reactive homeostasis is an immediate response to a homeostatic challenge such as predation."

However, *any* homeostasis is impossible without reaction - because homeostasis is and must be a "feedback" phenomenon.

The phrase "reactive homeostasis" is simply short for: "reactive compensation reestablishing homeostasis", that is to say, "reestablishing a point of homeostasis." - it should not be confused with a separate *kind* of homeostasis or a distinct phenomenon *from* homeostasis, it is simply the compensation (or compensatory) phase of homeostasis.

The term has come to be used in other fields, as well. An actuary may refer to *risk homeostasis*, where (for example) people who have anti-lock brakes have no better safety record than those without anti-lock brakes, because they unconsciously compensate for the safer vehicle via less-safe driving habits. Previously, certain maneuvers involved minor skids, evoking fear and avoidance: now the anti-lock system moves the boundary for such feedback, and behavior patterns expand into the no-longer punitive area. It has also been suggested that ecological crises are an instance of risk homeostasis in which behavior known to be dangerous continues until dramatic consequences actually occur. Sociologists and psychologists may refer to *stress homeostasis*, the tendency of a population or an individual to stay at a certain level of stress, often generating artificial stresses if the "natural" level of stress is not enough.

Jean Francois Lyotard, a postmodern theorist, has applied this term to societal 'power centers' that he describes as being 'governed by a principle of homeostasis.' For example the scientific hierarchy, which will sometimes ignore a radical new discovery for years because it destabilizes previously accepted norms.

Andrew Potter has used the term *waste homeostasis* in reference to the lack of net gain from energy saving technologies. A 2007 study purported to find (and show clinically) *conversational homeostasis* in which overly-familiar people (such as spouses) condense their speech so much that they are actually worse at communicating novel information than strangers are; while not being conscious of this problem.

Some herbal medicines, known as adaptogens, have been defined to function as non-toxic metabolic regulators that can enhance metabolic homeostasis during stress.

Evolution

A central organizing concept in biology is that all life has a common origin and has changed and developed through the process of the theory of evolution. This has led to the striking similarity of units and processes discussed in the previous section. Charles Darwin established evolution as a viable theory by articulating its driving force, natural selection (Alfred Russel Wallace is recognized as the co-discoverer of this concept). Darwin theorized that species and breeds developed through the processes of natural selection as well as by artificial selection or selective breeding.Genetic drift was embraced as an additional mechanism of evolutionary development in the modern synthesis of the theory.

The evolutionary history of a species— which describes the characteristics of the various species from which it descended— together with its genealogical relationship to every other species is called its phylogeny. Widely varied approaches to biology generate information about phylogeny. These include the comparisons of DNA sequences conducted within molecular biology or genomics, and comparisons of fossils or other records of ancient organisms in paleontology. Biologists organize and analyze evolutionary relationships through various methods, including phylogenetics, phenetics, and cladistics (The major events in the evolution of life, as biologists currently understand them, are summarized on this evolutionary timeline).

Ever since its articulation by Darwin and Wallace, the theory of evolution by natural selection has come under attack by people who disagree with scientific findings or interpretations regarding the origins and diversity of life, generally favoring instead religious explanations. See Creation-evolution controversy for more information.

Up into the 19th century, it was commonly believed that life forms could appear spontaneously under certain conditions. This misconception was challenged by William Harvey's diction that "all life [is] from [an] egg" (from the Latin "Omne vivum ex ovo"), a foundational concept of modern biology. It simply means that there is an unbroken continuity of life from its initial origin to the present time.

A group of organisms shares a common descent if they share a common ancestor. All organisms on the Earth have been and are descended from a common ancestor or an ancestral gene pool. This last universal common ancestor of all organisms is believed to have appeared about 3.5 billion years ago. Biologists generally regard the universality of the genetic code as definitive evidence in favor of the theory of universal common descent (UCD) for all bacteria, archaea, and eukaryotes.

Cell Theory

The generally accepted parts of cell theory include:

- The cell is the fundamental unit of structure and function in living things.

- All organisms are made up of one or more cells.
- All cells come from cells during cellular division.
- All cells are essentially the same in chemical composition.
- Energy flow (metabolism and biochemistry) occurs within cells.

This theory also contains two exceptions:

Viruses are considered by some to be alive, yet they are not made up of cells.

The first cell did not originate from a preexisting cell.

Essentially, cell theory states that all living things are composed of one or more cells (and products of those cells, for example, plasma, saliva, and other substances). Furthermore, those cells arise from other cells through cellular division, all cells in all organisms are similar in chemical composition, and that energy flow or metabolism occurs within the cells.

Gene Theory

Schematic representation of DNA, the primary genetic material. While organisms may vary immensely in appearance, habitat, and behaviour it is a central principle of biology that all life shares certain universal fundamentals.

A key feature is reproduction or replication. The entity being replicated, the replicator, in the past was considered to be the organism during the time of Darwin, but since the 1970s increasingly reduced to the scale of molecules. All known life has a carbon-based biochemistry, carbon is the fundamental building block of the molecules that make up all known living things.

Similarly water is the basic solvent for all known living organisms. While all these things are true of all organisms observed on Earth, in theory alternative forms of life could exist and some scientists do look at alternative biochemistry. All terrestrial organisms use DNA and RNA-based genetic mechanisms to hold genetic information. Another universal principle is that all observed organisms with the exception of viruses are made of cells. Similarly, all organisms share common developmental processes.

BIOGEOGRAPHY

Biogeography is the study of the distribution of biodiversity over space and time. It aims to reveal where organisms live, at what abundance, and why.

The patterns of species distribution at this level can usually be explained through a combination of historical factors such as speciation, extinction, continental drift, glaciation (and associated variations in sea level, river routes, and so on), and river capture, in combination with the area and isolation of landmasses (geographic constraints) and available energy supplies.

History

The theory of biogeography grows out of the work of Alfred Russel Wallace and other early evolutionary scientists. Wallace studied the distribution of flora and fauna in the Malay Archipelago in the 19th century. With the exception of Wallace and a few others, prior to the publication of *The Theory of Island Biogeography* by Robert MacArthur and E.O. Wilson in 1967 (which expanded their 1963 paper on the same topic) the field of biogeography was seen as a primarily historical one, and as such the field was seen as a purely descriptive one.

MacArthur and Wilson changed this perception, and showed that the species richness of an area could be predicted in terms of such factors as habitat area, immigration rate and extinction rate. This gave rise to an interest in island biogeography. The application of island biogeography theory to habitat fragments spurred the development of the fields of conservation biology and landscape ecology (at least among British and American academics; landscape ecology has a distinct genesis among European academics).

Classic biogeography has been expanded by the development of molecular systematics, creating a new discipline known as phylogeography. This development allowed scientists to test theories about the origin and dispersal of populations, such as island endemics. For example, while classic biogeographers were able to speculate about the origins of species in the Hawaiian Islands, phylogeography allows them to test theories of relatedness between these populations and putative source populations in Asia and North America.

Paleobiogeography goes one step further to include paleogeographic data and considerations of plate tectonics. Using molecular analyses and corroborated by fossils, it has been possible to demonstrate that perching birds evolved first in the region of Australia or the adjacent Antarctic (which at that time lay somewhat further north and had a temperate climate). From there, they spread to the other Gondwanan continents and Southeast Asia - the part of Laurasia then closest to their origin of dispersal - in the late Paleogene, before achieving a global distribution in the early Neogene (Jønsson & Fjeldså 2006). Not knowing the fact that at the time of dispersal, the Indian Ocean was much narrower than it is today, and that South America was closer to the Antarctic, one would be hard pressed to explain the presence of many "ancient" lineages of perching birds in Africa, as well as the mainly South American distribution of the suboscines.

Classification

Biogeography is a synthetic science, related to geography, biology, soil science, geology, climatology, ecology and evolution.

Some fundamentals in biogeography are

- evolution (change in genetic composition of a population)
- extinction (disappearance of a species)
- dispersal (movement of populations away from their point of origin, related to migration
- range and distribution
- endemic areas
- vicariance

FRESHWATER BIOLOGY

Freshwater biology is a field of biology that studies the life and ecosystems of freshwater habitats. This includes animal, plant and microbial life in lakes, rivers and ponds. Fresh water habitats can be classified by temperature, light penetration, and the vegetation.

1. Limnology

Limnology (from Greek: *limne*, "lake"; and ëüãïò, *logos*, "knowledge") is the study of inland waters (fresh), including their biological, physical, chemical, geological and hydrological aspects. This includes the study of (natural and man-made) lakes and ponds, rivers and streams, wetlands and groundwaters.

François-Alphonse Forel (1841-1912) established the field with his studies of Lake Geneva. Limnology traditionally is closely related to hydrobiology, which is concerned with the application of the principles and methods of physics, chemistry, geology, and geography to ecological problems.

Main Organizations

- American Society of Limnology and Oceanography
- Australian Society for Limnology
- European Society of Limnology and Oceanography
- German Society of Limnology
- Italian Association for Oceanology and Limnology (AIOL)

- The Japanese Society of Limnology
- Societas Internationalis Limnologiae (SIL) (limnology.org)
- Brazilian Society of Limnology
- New Zealand freshwater Sciences society
- Southern African Society of Aquatic Scientists

2. Marine Biology

Marine biology is the scientific study of the plants, animals and other organisms that live in the ocean or any other body of water. Given that in biology many phyla, families and genera have some species that live in the sea and others that live on land, marine biology classifies species based on the environment rather than on taxonomy.

There are many practical reasons to study marine biology. Marine life represents a vast resource, providing food, medicine, and raw materials, in addition to helping to support recreation and tourism all over the world. At a fundamental level, marine life helps determine the very nature of our planet. Marine organisms produce much of the oxygen we breathe and probably help regulate the earth's climate. Shorelines are in part shaped and protected by marine life, and some marine organisms even help create new land.

Overview

Marine biology covers a great deal, from the microscopic, including plankton and phytoplankton, which can be as small as 0.02 micrometers and are both hugely important as the primary producers of the sea, to the huge cetaceans (whales) which reach up to a reported 33 meters (109 feet) in length.

The habitats studied by marine biology include everything from the tiny layers of surface water in which organisms and abiotic items may be trapped in surface tension between the ocean and atmosphere, to the depths of the abyssal trenches, sometimes 10,000 meters or more beneath the surface of the ocean. It studies habitats such as coral reefs, kelp forests, tidepools, muddy, sandy, and rocky bottoms, and the open ocean (pelagic) zone, where solid objects are rare and the surface of the water is the only visible boundary.

A large amount of all life on Earth exists in the oceans. Exactly how large the proportion is is still unknown. While the oceans comprise about 71% of the Earth's surface, due to their depth they encompass about 300 times the habitable volume of the terrestrial habitats on Earth.

Many species are economically important to humans, including the food fishes. It is also becoming understood that the well-being of marine organisms and other organisms are linked in very fundamental ways. Human understanding is growing of the relationship between life in the sea and important cycles such as that of matter (such as the carbon cycle) and of air (such as Earth's respiration, and movement of energy through ecosystems). Large areas beneath the ocean surface still remain effectively unexplored. Scientists know more about the moon than they know about the ocean and the life in it.

Subfields

The marine ecosystem is large, and thus there are many subfields of marine biology. Most involve studying specializations of particular species (*i.e.*, phycology, invertebrate zoology, ichthyology).

Other subfields study the physical effects of continual immersion in sea water and the ocean in general, adaptation to a salty environment, and the affects of changing various oceanic properties on marine life. A subfield of marine biology studies the relationships between oceans and ocean life, and global weather and environmental issues (such as carbon dioxide displacement).

Recent marine biotechnology has focused largely on marine biomolecules, especially proteins, that may have uses in medicine or engineering. Marine environments are the home to many exotic biological materials that may inspire biomimetic materials.

An interesting branch of marine biology is aquaculture; which some countries do a lot of in the oceans, especially Japan.

Related Fields

Marine biology is closely linked to both oceanography and biology. It also encompasses many ideas from ecology. Fisheries science and marine conservation can be considered partial offshoots of marine biology.

Lifeforms

Microscopic Life

Microscopic life undersea is incredibly diverse and still poorly understood. For example, the role of viruses in marine ecosystems is barely being explored even in the beginning of the 21st century.

The role of phytoplankton is better understood due to their critical position as the most numerous primary producers on Earth. Phytoplankton are categorized into cyanobacteria (also called blue-green algae/bacteria), various types of algae (red, green, brown, and yellow-green), diatoms, dinoflagellates, euglenoids, coccolithophorids, cryptomonads, chrysophytes, chlorophytes, prasinophytes, and silicoflagellates.

Zooplankton tend to be somewhat larger, and not all are microscopic. Many Protozoa are zooplankton, including dinoflagellates, zooflagellates, foraminiferans, and radiolarians. Some of these (such as dinoflaggelates) are also phytoplankton; the plant/animal distinction often breaks down in very small organisms. Other zooplankton include cnidarians, ctenophores, chaetognaths, molluscs, arthropods, urochordates, and annelids such as polychaetes. Many larger animals begin their life as zooplankton before they become large enough to take their familiar forms. Two examples are fish larvae and sea stars (also called starfish).

Plants and Algae

Plant life is relatively rare undersea. Most of the niche occupied by plants on land is actually occupied by macroscopic algae in the ocean, such as *Sargassum* and kelp which are commonly known as seaweeds. The non algae plants that do survive in the sea are often found in shallow waters, such as the seagrasses (examples of which are eelgrass, *Zostera*, and turtlegrass, *Thalassia*). These plants have adapted to the high salinity of the ocean environment. The intertidal zone is also a good place to find plant life in the sea, where mangroves or cordgrass or beach grass might grow.

Marine Invertebrates

As on land, invertebrates make up a huge portion of all life in the sea. Invertrebrate sea life includes Cnidaria such as Jellyfish and sea anemone; Ctenophora; sea worms including phyla: Plathyhelminthes, Nemertea, Annelida, Sipuncula, Echiura, and the Phoronida; Mollusca including shellfish, squid, octopus; Crustaceans; Porifera including sponges, Bryozoa, Echinodermata including starfish; and Urochordata - sea squirts or tunicates.

Fish

Fish have evolved very different biological functions from other large organisms. Fish anatomy includes a two-chambered heart, operculum, secretory cells that produce mucous, swim bladder, scales, fins, lips and eyes. Fish breathe by extracting oxygen from water through their gills. Fins propel and stabilize the fish in the water.

Well known fish include: sardines, anchovy, tuna, clownfish (also known as anemonefish), and bottom fish which include halibut and ling cod. Predators include sharks and barracuda.

Reptiles

Reptiles which inhabit or frequent the sea include sea turtles, marine iguanas, sea snakes, and saltwater crocodiles. Most reptile species live on or near land, rather than in the ocean.

Seabirds

Seabirds are species of [Bird]]s adapted to living in the marine environment, examples including albatross, penguins, gannets, and auks. Although they spend most of their lives in the ocean, species such as gulls can often be found thousands of miles inland.

Marine Mammals

There are five main types of marine mammals.

- Cetaceans include toothed whales (Suborder Odontoceti), such as the Sperm Whale, dolphins, and porpoises. Cetaceans also include baleen whales (Suborder Mysticeti), such as Gray Whales, Humpback Whales, Dall's porpoise, and Blue Whales.
- Sirenians include manatees, the Dugong, and the extinct Steller's Sea Cow.
- Seals (Family Phocidae), sea lions (Family Otariidae - which also include the fur seals), and the Walrus (Family Odobenidae) are all considered pinnipeds.
- Sea Otters are members of the Family Mustelidae, which includes weasels and badgers.
- Finally, Polar Bears (Family Ursidae) are sometimes considered marine mammals because of their dependence on the sea.

Oceanic Habitats

Reefs

Reefs comprise some of the densest and most diverse habitats in the world. The best-known types of reefs are tropical coral reefs which exist in most tropical waters; however, reefs can also exist in cold water. Reefs are built up by corals and other calcium-depositing animals, usually on top of a rocky outcrop on the ocean floor. Reefs can also grow on other surfaces, which has made it possible to create artificial reefs. Coral reefs also support a huge community of life, including the corals themselves, their symbiotic zooxanthellae, tropical fish and many other organisms.

Much attention in marine biology is focused on coral reefs and the El Niño weather phenomenon. In 1998, coral reefs experienced a "once in a thousand years" bleaching event, in which vast expanses of reefs across the Earth died because sea surface temperatures rose well above normal. Some reefs are recovering, but scientists say that 58% of the world's coral reefs are now endangered and predict that global warming could exacerbate this trend.

Deep Sea and Trenches

The ocean is deep, very deep in some places. The deepest recorded measure to date is the Mariana Trench, near the Philippines, in the Pacific Ocean at 10924 m (35838 ft). At such depths, water pressure is extreme and there is no sunlight, but some life still exists. Small flounder (family Soleidae) fish and shrimp were seen by the American crew of the bathyscaphe *Trieste* when it dove to the bottom in 1960.

Other notable oceanic trenches include Monterey Canyon, in the eastern Pacific, the Tonga Trench in the southwest at 10,882 m (35,702 feet), the Philippine Trench, the Puerto Rico Trench at 8605 m (28232 ft), the Romanche Trench at 7760 m (24450 ft), Fram Basin in the Arctic Ocean at 4665 m (15305 ft), the Java Trench at 7450 m (24442 ft), and the South Sandwich Trench at 7235 m (23737 ft).

In general, the deep sea is considered to start at the aphotic zone, the point where sunlight loses its power of transference through the water. Many life forms that live at these depths have the ability to create their own light.

Much life centers on seamounts that rise from the deeps, where fish and other sea life congregate to spawn and feed. Hydrothermal vents along the mid-ocean ridge spreading centers act as oases, as do their opposites, cold seeps. Such places support unique biomes and many new microbes and other lifeforms have been discovered at these locations.

Open Ocean

The great expanse of open ocean habitat is huge, and many species can be found passing through it and living in it. The term "open ocean" usually is meant to refer to the vast stretches of water between points of land, or between undersea mounts. Contrary to popular notions the open ocean is often not the place where marine animals spend the majority of their lives. Most species simply pass through the open ocean on their ways to other places. Larger species are the main ongoing inhabitants.

Intertidal and Shore

Intertidal zones, those areas close to shore, are constantly being exposed and covered by the ocean's tides. A huge array of life lives within this zone.

Shore habitats span from the upper intertidal zones to the area where land vegetation takes prominence. It can be underwater anwhere from daily to very infrequently. Many species here are scavengers, living off of sea life that is washed up on the shore. Many land animals also make much use of the shore and intertidal habitats. A subgroup of organisms in this habitat bores and grinds exposed rock through the process of bioerosion.

How Oceanic Factors Affect Distribution of Various Organisms

An active research topic in marine biology is to discover and map the life cycles of various species and where they spend their time. Marine biologists study how the ocean currents, tides and many other oceanic factors affect ocean lifeforms, including their growth, distribution and well-being. This has only recently become technically feasible with advances in GPS and newer underwater visual devices.

Most ocean life breeds in specific places, nests or not in others, spends time as juveniles in still others, and in maturity in yet others. Scientists know little about where many species spent different parts of their life cycles. For example, it is still largely unknown where sea turtles travel. Tracking devices do not work for some life forms, and the ocean is not friendly to technology. But these factors are being

Famous Marine Biologists

- Ali Abdelghany (1944-) Egyptian marine biologist
- Jakob Johan Adolf Appellöf (1857-1921), Swedish marine zoologist.
- Joseph Ayers, marine neurophysiologist and biomimetic researcher.
- Samuel Stillman Berry (1887-1984), U.S. marine zoologist.
- Henry Bryant Bigelow (1879 – 1967), U.S. marine biologist.
- Rachel Carson (1907-1964) American Marine Biologist and Author.
- Ilham Artüz (1924-1993) Turkish Marine Biologist and Oceanographer
- Carl Chun (1852-1914), German marine biologist
- Jacques-Yves Cousteau (1910-1997), French marine biologist and explorer.
- Anton Dohrn (1840-1909), German marine biologist.
- Sylvia Earle (born 1935), American oceanographer.
- Hans Hass (born 1919), Austrian marine biologist and diving pioneer
- Gotthilf Hempel (born 1929), German marine biologist
- Johan Hjort, Norwegian marine zoologist and one of the founders of ICES.
- Bruno Hofer (1861-1916), German fisheries scientist.
- Hirohito (1901-1989), Japanese emperor and jellyfish taxonomist
- Uwe Kils (born 1951), German marine biologist
- August David Krohn (1803 – 1891), Russian/German zoologist.
- William Elford Leach (1790-1836), English zoologist and marine biologist.
- Philip S. Lobel Professor, Boston University Marine Program

- Nicholai Miklukho-Maklai (1846-1888), Russian marine biologist and anthropologist.
- Sir John Murray (1841-1914), Scots-Canadian marine biologist.
- Harald Rosenthal (born 1937), German hydrobiologist known for his work in fish farming and ecology.
- Michael Sars (1809 – 1869), Norwegian theologian and biologist.
- Georg Sars (1837-1927), Norwegian marine biologist.
- Ruth Turner (1915-2000), marine biologist.
- Robert Paine Professor, emeritus University of Washington intertidal Ecologist.
- Y(1832-1882), Scottish marine biologist.

REFERENCES

Dansereau, Pierre (1957): *Biogeography: An Ecological Perspective.* Ronald Press Company, New York City.

Jonsson, Knud A. & Fjeldsa, Jon (2006): *Determining biogeographical patterns of dispersal and diversification in oscine passerine birds in Australia*, Southeast Asia and Africa. J. Biogeogr. 33 (7): 1155–1165.

3

Origin of Life and Bioscience Timeline

ORIGIN OF LIFE

In the natural sciences, abiogenesis, the question of the origin of life, is the study of how life on Earth might have emerged from non-life. Scientific consensus is that abiogenesis occurred sometime between 4.4 billion years ago, when water vapor first liquefied, and 2.7 billion years ago, when the ratio of stable isotopes of carbon (12C and 13C), iron and sulfur points to a biogenic origin of minerals and sediments and molecular biomarkers indicate photosynthesis. This topic also includes panspermia and other exogenic theories regarding possible extra-planetary or extra-terrestrial origins of life, thought to have possibly occurred sometime over the last 13.7 billion years in the evolution of the Universe since the Big Bang.

Origin of life studies is a limited field of research despite its profound impact on biology and human understanding of the natural world. Progress in this field is generally slow and sporadic, though it still draws the attention of many due to the eminence of the question being investigated. Several theories have been proposed, most notably RNA world hypothesis. For the observed evolution of life on earth, see the timeline of life.

History of the Concept in Science

In the early Nineteenth Century and before people frequently believed that life arose spontaneously from non-living matter.

Darwin & Pasteur

By the middle of the 19th century Pasteur and others had demonstrated that living organisms did not arise spontaneously from non-living matter; the question therefore arose of how life might have come about within a naturalistic framework. In a letter to Joseph Dalton Hooker on February 1, 1871, Charles Darwin made the suggestion that the original spark of life may have begun in a "warm little pond, with all sorts of ammonia and phosphoric salts, lights, heat, electricity, etc. present, so that a protein compound was chemically formed ready to undergo still more complex changes". He went on to explain that "at the present day such matter would be instantly devoured or absorbed, which would not have been the case before living creatures were formed." In other words, the presence of life itself makes the search for the origin of life dependent on the sterile conditions of the laboratory.

Haldane & Oparin

No real progress was made until 1924 when Aleksandr Ivanovich Oparin experimentally showed that atmospheric oxygen prevented the synthesis of the organic molecules that are the necessary building blocks for the evolution of life. In his *The Origin of Life on Earth*, Oparin argued that a "primeval soup" of organic molecules could be created in an oxygen-less atmosphere through the action of sunlight. These would combine in ever-more complex fashions until they dissolved into a coacervate droplet. These droplets would "grow" by fusion with other droplets, and "reproduce" through fission into daughter droplets, and so have a primitive metabolism in which those factors which promote "cell integrity" survive, those that do not become extinct.

Many modern theories of the origin of life still take Oparin's ideas as a starting point. Around the same time J. B. S. Haldane also suggested that the earth's pre-biotic oceans - very different from their modern counterparts - would have formed a "hot dilute soup" in which organic compounds, the building blocks of life, could have formed. This idea was called biopoiesis or biopoesis, the process of living matter evolving from self-replicating but nonliving molecules.

Early Conditions

Morse and MacKenzie have suggested that oceans may have appeared first in the Hadean era, as soon as 200 million years after the Earth was formed, in a hot (100 °C) reducing environment, and that the pH of about 5.8 rose rapidly towards neutral. This has been supported by Wilde who has pushed the date of the zircon crystals found in the metamorphosed quartzite of Mount Narryer in Western Australia, previously thought to be 4.1-4.2 billion years old, to 4.404 billion years. This means that oceans and continental crust existed within 150 million years of Earth's formation. Despite this, the Hadean environment was one highly hazardous to life. Frequent collisions with large objects, up to 500 kilometres in diameter, would have been sufficient to vaporise the ocean within a few months of impact, with hot steam mixed with rock vapour leading to high altitude clouds completely covering the planet. After a few months the height of these clouds begins to fall but the cloud base would still be elevated probably for the next thousand years after which at low altitude it starts to rain. For another two thousand years rains slowly draw down the height of the clouds, returning the oceans to their original depth only 3,000 years after the impact event. The possible Late Heavy Bombardment possibly caused by the movements in position of the Gaseous Giant planets, that pockmarked the moon, and other inner planets (Mercury, Mars, and presumably Earth and Venus), between 3.8 and 4.1 billion years would likely have sterilised the planet if life had already evolved by that time.

Evidence of the early appearance of life comes from the Isua supercrustal belt in Western Greenland and from similar formations in nearby the Akilia Islands. Carbon entering into rock formations has a concentration of elemental äC13 of about -5.5, where because of a preferential biotic uptake of C12, biomass has a äC13 of between -20 and -30. These isotopic fingerprints are preserved in the sediments, and Mojzis has used this technique to suggest that life existed on the planet already by 3.85 billion years ago. The rapidity of the evolution of life Lazcano and Miller (1994) suggest, is dictated by the fact that all of the ocean's waters is recirculated through mid ocean submarine vents every 10 million years, and any organic compounds produced by then would be altered or destroyed by temperatures exceeding 300 °C. They estimate that the development of a 100 kilobase genome of a DNA/protein primative heterotroph into a 7000 gene filamentous cyanobacterium would have required only 7 million years.

Current Models

There is no truly "standard model" of the origin of life. But most currently accepted models build in one way or another upon a number of discoveries about the origin of molecular and cellular components for life, which are listed in a rough order of postulated emergence:

1. Plausible pre-biotic conditions result in the creation of certain basic small molecules (monomers) of life, such as amino acids. This was demonstrated in the Miller-Urey experiment by Stanley L. Miller and Harold C. Urey in 1953.
2. Phospholipids (of an appropriate length) can spontaneously form lipid bilayers, a basic component of the cell membrane.
3. The polymerization of nucleotides into random RNA molecules might have resulted in self-replicating *ribozymes* (*RNA world hypothesis*).

4. Selection pressures for catalytic efficiency and diversity result in ribozymes which catalyse peptidyl transfer (hence formation of small proteins), since oligopeptides complex with RNA to form better catalysts. Thus the first ribosome is born, and protein synthesis becomes more prevalent.

5. Proteins outcompete ribozymes in catalytic ability, and therefore become the dominant biopolymer. Nucleic acids are restricted to predominantly genomic use.

The origin of the basic biomolecules, while not settled, is less controversial than the significance and order of steps 2 and 3. The basic chemicals from which life was thought to have formed are:

- methane (CH_4),
- ammonia (NH_3),
- water (H_2O),
- hydrogen sulfide (H_2S),
- carbon dioxide (CO_2) or carbon monoxide (CO), and
- phosphate (PO_{43}^-).

Molecular oxygen (O_2) and ozone (O_3) were either rare or absent.

As oi 2007, no one has yet synthesized a "protocell" using basic components which would have the necessary properties of life (the so-called *"bottom-up-approach"*). Without such a proof-of-principle, explanations have tended to be short on specifics. However, some researchers are working in this field, notably Steen Rasmussen at Los Alamos National Laboratory and Jack Szostak at Harvard University. Others have argued that a *"top-down approach"* is more feasible. One such approach, attempted by Craig Venter and others at The Institute for Genomic Research, involves engineering existing prokaryotic cells with progressively fewer genes, attempting to discern at which point the most minimal requirements for life were reached. The biologist John Desmond Bernal, coined the term Biopoesis for this process, and suggested that there were a number of clearly defined "stages" that could be recognised in explaining the origin of life.

- Stage 1: The origin of biological monomers
- Stage 2: The origin of biological polymers
- Stage 3: The evolution from molecules to cell

Bernal suggested that Darwinian evolution may have commenced early, some time between Stage 1 and 2.

Origin of Organic Molecules

Miller's Experiments

In 1953 a graduate student, Stanley Miller, and his professor, Harold Urey, performed an experiment that proved organic molecules could have spontaneously formed on Early Earth from inorganic precursors. The now-famous "Miller-Urey experiment" used a highly reduced mixture of gases - methane, ammonia and hydrogen – to form basic organic monomers, such as amino acids. Whether the mixture of gases used in the Miller-Urey experiment truly reflects the atmospheric content of Early Earth is a controversial topic. Other less reducing gases produce a lower yield and variety. It was once thought that appreciable amounts of molecular oxygen were present in the prebiotic atmosphere, which would have essentially prevented the formation of organic molecules; however, the current scientific consensus is that such was not the case.

Simple organic molecules are, of course, a long way from a fully functional self-replicating life form. But in an environment with no pre-existing life these molecules may have accumulated and provided a rich environment for chemical evolution ("soup theory"). On the other hand, the spontaneous formation of complex polymers from abiotically generated monomers under these conditions is not at all a straightforward process. Besides the necessary basic organic monomers, compounds that would have prohibited the formation of polymers were formed in high concentration during the experiments. It can be argued that the most crucial challenge unanswered by this theory is how the relatively simple organic building blocks polymerise and form more complex structures, interacting in consistent ways to form a protocell. For example, in an aqueous environment hydrolysis of oligomers/polymers into their constituent monomers would be favored over the condensation of individual monomers into polymers. Also, the Miller experiment produces many substances that would undergo cross-reactions with the amino acids or terminate the peptide chain.

Fox's Experiments

In the 1950s and 1960s Sidney W. Fox, studied the spontaneous formation of peptide structures under conditions that might plausibly have existed early in Earth's history. He demonstrated that amino acids could spontaneously form small peptides. These amino acids and small peptides could be encouraged to form closed spherical membranes, called microspheres. Fox described these formations as protocells, protein spheres that could grow and reproduce.

Eigen's Hypothesis

In the early 1970s the problem of the origin of life was approached by Manfred

Eigen and Peter Schuster of the Max Planck Institute for Biophysical Chemistry. They examined the transient stages between the molecular chaos and a self replicating hypercycle in a prebiotic soup.

In a hypercycle, the information storing system (possibly RNA) produces an enzyme, which catalyzes the formation of another information system, in sequence until the product of the last aids in the formation of the first information system. Mathematically treated, hypercycles could create quasispecies, which through natural selection entered into a form of Darwinian evolution. A boost to hypercycle theory was the discovery that RNA, in certain circumstances forms itself into ribozymes, a form of RNA enzyme.

Wächtershäuser's Hypothesis

Another possible answer to this polymerization conundrum was provided in 1980s by Günter Wächtershäuser, in his iron-sulfur world theory. In this theory, he postulated the evolution of (bio)chemical pathways as fundamentals of the evolution of life. Moreover, he presented a consistent system of tracing today's biochemistry back to ancestral reactions that provide alternative pathways to the synthesis of organic building blocks from simple gaseous compounds.

In contrast to the classical Miller experiments, which depend on external sources of energy (such as simulated lightning or UV irradiation), "Wächtershäuser systems" come with a built-in source of energy, sulfides of iron and other minerals (e.g. pyrite). The energy released from redox reactions of these metal sulfides is not only available for the synthesis of organic molecules, but also for the formation of oligomers and polymers. It is therefore hypothesized that such systems may be able to evolve into autocatalytic sets of self-replicating, metabolically active entities that would predate the life forms known today.

The experiment produced a relatively small yield of dipeptides (0.4% to 12.4%) and a smaller yield of tripeptides (0.10%) but the authors also noted that: "under these same conditions dipeptides hydrolysed rapidly." Another criticism of the result is that the experiment did not include any organic molecules that would most likely cross-react or chain-terminate.

William Martin and Michael Russell reported a modified iron-sulfur-hypothesis in 2002. According to their scenario, the first cellular life forms may have evolved inside so-called black smokers at seafloor spreading zones in the deep sea. These structures consist of microscale caverns that are coated by thin membraneous metal sulfide walls. Therefore, these structures would solve several critical points of the "pure" Wächtershäuser systems at once:

1. the micro-caverns provide a means of concentrating newly synthesised

molecules, thereby increasing the chance of forming oligomers;

2. the steep temperature gradients inside a black smoker allow for establishing "optimum zones" of partial reactions in different regions of the black smoker (e.g. monomer synthesis in the hotter, oligomerisation in the colder parts);
3. the flow of hydrothermal water through the structure provides a constant source of building blocks and energy (freshly precipitated metal sulfides);
4. the model allows for a succession of different steps of cellular evolution (prebiotic chemistry, monomer and oligomer synthesis, peptide and protein synthesis, RNA world, ribonucleoprotein assembly and DNA world) in a single structure, facilitating exchange between all developmental stages;
5. synthesis of lipids as a means of "closing" the cells against the environment is not necessary, until basically all cellular functions are developed.

This model locates the "last universal common ancestor" (LUCA) inside a black smoker, rather than assuming the existence of a free-living form of LUCA. The last evolutionary step would be the synthesis of a lipid membrane that finally allows the organisms to leave the microcavern system of the black smokers and start their independent lives. This postulated late acquisition of lipids is consistent with the presence of completely different types of membrane lipids in archaebacteria and eubacteria (plus eukaryotes) with highly similar cellular physiology of all life forms in most other aspects.

From Organic Molecules to Protocells

The question "How do simple organic molecules form a protocell?" is largely unanswered but there are many hypotheses. Some of these postulate the early appearance of nucleic acids ("genes-first") whereas others postulate the evolution of biochemical reactions and pathways first ("metabolism-first"). Recently, trends are emerging to create hybrid models that combine aspects of both.

"Genes First" Models: The RNA World

The RNA world hypothesis suggests that relatively short RNA molecules could have spontaneously formed that were capable of catalyzing their own continuing replication. It is difficult to gauge the probability of this formation. A number of theories of modes of formation have been put forward. Early cell membranes could have formed spontaneously from proteinoids, protein-like molecules that are produced when amino acid solutions are heated - when present at the correct concentration in aqueous solution, these form microspheres which are observed to behave similarly to membrane-enclosed compartments. Other possibilities include systems of chemical reactions taking place within clay substrates or on the surface of pyrite rocks.

Factors supportive of an important role for RNA in early life include its ability to act both to store information and catalyse chemical reactions (as a ribozyme); its many important roles as an intermediate in the expression and maintenance of the genetic information (in the form of DNA) in modern organisms; and the ease of chemical synthesis of at least the components of the molecule under conditions approximating the early Earth. Relatively short RNA molecules which can duplicate others have been artificially produced in the lab.

A slightly different version of this hypothesis is that a different type of nucleic acid, such as PNA or TNA, was the first one to emerge as a self-reproducing molecule, to be replaced by RNA only later.

"Metabolism First" Models: Iron-Sulfur World and Others

Several models reject the idea of the self-replication of a "naked-gene" and postulate the emergence of a primitive metabolism which could provide an environment for the later emergence of RNA replication.

One of the earliest incarnations of this idea was put forward in 1924 with Aleksandr Ivanovich Oparin's notion of primitive self-replicating vesicles which predated the discovery of the structure of DNA. More recent variants in the 1980s and 1990s include Günter Wächtershäuser's iron-sulfur world theory and models introduced by Christian de Duve based on the chemistry of thioesters. More abstract and theoretical arguments for the plausibility of the emergence of metabolism without the presence of genes include a mathematical model introduced by Freeman Dyson in the early 1980s and Stuart Kauffman's notion of collectively autocatalytic sets, discussed later in that decade.

However, the idea that a closed metabolic cycle, such as the reductive citric acid cycle, could form spontaneously (proposed by Günter Wächtershäuser) remains unsupported. According to Leslie Orgel, a leader in origin-of-life studies for the past several decades, there is reason to believe the assertion will remain so. In an article entitled "Self-Organizing Biochemical Cycles", Orgel summarizes his analysis of the proposal by stating, "There is at present no reason to expect that multistep cycles such as the reductive citric acid cycle will self-organize on the surface of FeS/FeS2 or some other mineral." It is possible that another type of metabolic pathway was used at the beginning of life. For example, instead of the reductive citric acid cycle, the "open" acetyl-CoA pathway (another one of the four recognised ways of carbon dioxide fixation in nature today) would be even more compatible with the idea of self-organisation on a metal sulfide surface. The key enzyme of this pathway, carbon monoxide dehydrogenase/acetyl-CoA synthase harbours mixed nickel-iron-sulfur clusters in its reaction centers and catalyses the formation of acetyl-CoA (which may be regarded as a modern form of acetyl-thiol) in a single step.

Bubble Theory

Waves breaking on the shore create a delicate foam composed of bubbles. Winds sweeping across the ocean have a tendency to drive things to shore, much like driftwood collecting on the beach. It is possible that organic molecules were concentrated on the shorelines in much the same way. Shallow coastal waters also tend to be warmer, further concentrating the molecules through evaporation. While bubbles composed mostly of water burst quickly, water containing amphiphiles forms much more stable bubbles, lending more time to the particular bubble to perform these crucial experiments.

Amphiphiles are oily compounds containing a hydrophilic head on one or both ends of a hydrophobic molecule. Some amphiphiles have the tendency to spontaneously form membranes in water. A spherically closed membrane contains water and is a hypothetical precursor to the modern cell membrane. If a protein came along that increased the integrity of its parent bubble, then that bubble had an advantage, and was placed at the top of the natural selection waiting list. Primitive reproduction can be envisioned when the bubbles burst, releasing the results of the experiment into the surrounding medium. Once enough of the 'right stuff' was released into the medium, the development of the first prokaryotes, eukaryotes, and multicellular organisms could be achieved.

Similarly, bubbles formed entirely out of protein-like molecules, called microspheres, will form spontaneously under the right conditions. But they are not a likely precursor to the modern cell membrane, as cell membranes are composed primarily of lipid compounds rather than amino-acid compounds (for types of membrane spheres associated with abiogenesis, see protobionts, micelle, coacervate).

A recent model by Fernando and Rowe suggests that the enclosure of an autocatalytic non-enzymatic metabolism within protocells may have been one way of avoiding the side-reaction problem that is typical of metabolism first models.

Other Models

Autocatalysis

British ethologist Richard Dawkins wrote about autocatalysis as a potential explanation for the origin of life in his 2004 book *The Ancestor's Tale*. Autocatalysts are substances which catalyze the production of themselves, and therefore have the property of being a simple molecular replicator. In his book, Dawkins cites experiments performed by Julius Rebek and his colleagues at the Scripps Research Institute in California in which they combined amino adenosine and pentafluorophenyl ester with the autocatalyst amino adenosine triacid ester (AATE). One system from the experiment contained variants of AATE which catalysed the synthesis of themselves.

This experiment demonstrated the possibility that autocatalysts could exhibit competition within a population of entities with heredity, which could be interpreted as a rudimentary form of natural selection.

Clay Theory

A model for the origin of life based on clay was forwarded by Dr A. Graham Cairns-Smith of the University of Glasgow in 1985 and adopted as a plausible illustration by several other scientists, including Richard Dawkins. Clay theory postulates that complex organic molecules arose gradually on a pre-existing, non-organic replication platform — silicate crystals in solution. Complexity in companion molecules developed as a function of selection pressures on types of clay crystal is then exapted to serve the replication of organic molecules independently of their silicate "launch stage". It is, truly, "life from a rock."

Cairns-Smith is a staunch critic of other models of chemical evolution. However, he admits, that like many models of the origin of life, his own also has its shortcomings (Horgan 1991).

In 2007, Kahr and colleagues reported their experiments to examine the idea that crystals can act as a source of transferable information, using crystals of potassium hydrogen phthalate. "Mother" crystals with imperfections were cleaved and used as seeds to grow "daughter" crystals from solution. They then examined the distribution of imperfections in the crystal system and found that the imperfections in the mother crystals were indeed reproduced in the daughters. The daughter crystals had many additional imperfections. For a gene-like behavior the additional imperfections should be much less than the parent ones, thus Kahr concludes that the crystals "were not faithful enough to store and transfer information form one generation to the next".

"Deep-Hot Biosphere" Model of Gold

The discovery of nanobes (filamental structures that are smaller than bacteria, but that may contain DNA) in deep rocks, led to a controversial theory put forward by Thomas Gold in the 1990s that life first developed not on the surface of the Earth, but several kilometers below the surface. It is now known that microbial life is plentiful up to five kilometers below the earth's surface in the form of archaea, which are generally considered to have originated either before or around the same time as eubacteria, most of which live on the surface including the oceans. It is claimed that discovery of microbial life below the surface of another body in our solar system would lend significant credence to this theory. He also noted that a trickle of food from a deep, unreachable, source is needed for survival because life arising in a puddle of organic material is likely to consume all of its food and become extinct.

"Primitive" Extraterrestrial Life

An alternative to Earthly abiogenesis is the hypothesis that primitive life may have originally formed extraterrestrially, either in space or on a nearby planet (Mars). (Note that *exogenesis* is related to, but not the same as, the notion of panspermia). A supporter of this theory is Francis Crick.

Organic compounds are relatively common in space, especially in the outer solar system where volatiles are not evaporated by solar heating. Comets are encrusted by outer layers of dark material, thought to be a tar-like substance composed of complex organic material formed from simple carbon compounds after reactions initiated mostly by irradiation by ultraviolet light. It is supposed that a rain of material from comets could have brought significant quantities of such complex organic molecules to Earth.

An alternative but related hypothesis, proposed to explain the presence of life on Earth so soon after the planet had cooled down, with apparently very little time for prebiotic evolution, is that life formed first on early Mars. Due to its smaller size Mars cooled before Earth (a difference of hundreds of millions of years), allowing prebiotic processes there while Earth was still too hot. Life was then transported to the cooled Earth when crustal material was blasted off Mars by asteroid and comet impacts. Mars continued to cool faster and eventually became hostile to the continued evolution or even existence of life (it lost its atmosphere due to low volcanism), Earth is following the same fate as Mars, but at a slower rate.

Neither hypothesis actually answers the question of how life first originated, but merely shifts it to another planet or a comet. However, the advantage of an extraterrestrial origin of primitive life is that life is not required to have evolved on each planet it occurs on, but rather in a single location, and then spread about the galaxy to other star systems via cometary and/or meteorite impact. Evidence to support the plausibility of the concept is scant, but it finds support in recent study of Martian meteorites found in Antarctica and in studies of extremophile microbes. Additional support comes from a recent discovery of a bacterial ecosytem whose energy source is radioactivity.

The Lipid World

There is a theory that ascribes the first self-replicating object to be lipid-like. It is known that phospholipids spontaneously form bilayers in water - the same structure as in cell membranes. These molecules were not present on early earth, however other amphiphilic long chain molecules also form membranes. Furthermore, these bodies may expand (by insertion of additional lipids), and under excessive expansion may undergo spontaneous splitting which preserves the same size and composition of lipids in the two progenies. The main idea in this theory is that the molecular composition of the lipid bodies is the preliminary way for information storage, and evolution led to the appearance of polymer entities such as RNA or DNA that may store information favorably.

The Polyphosphate Model

The problem with most scenarios of abiogenesis is that the thermodynamic equilibrium of amino acid versus peptides is in the direction of separate amino acids. What has been missing is some force that drives polymerization. The resolution of this problem may well be in the properties of polyphosphates. Polyphosphates are formed by polymerization of ordinary monophosphate ions PO4-3 by ultraviolet light. Polyphosphates cause polymerization of amino acids into peptides. Ample ultraviolet light must have existed in the early oceans. The key issue seems to be that calcium reacts with soluble phosphate to form insoluble calcium phosphate (apatite), so some plausible mechanism must be found to keep free calcium ions from solution. Possibly, the answer may be in some stable, non-reactive complex such as calcium citrate.

The Ecopoesis Model

The Ecopoesis model proposes that the geochemical cycles of biogenic elements, driven by an early oxygen-rich atmosphere, were the basis of a planetary metabolism that preceded and conditioned the gradual evolution of organismal life.

PAH World Hypothesis

Other sources of complex molecules have been postulated, including extra-terrestrial stellar or interstellar origin. For example, from spectral analyses, organic molecules are known to be present in comets and meteorites. In 2004, a team detected traces of polycyclic aromatic hydrocarbons (PAH's) in a nebula. Those are the most complex molecules so far found in space. The use of PAH's has also been proposed as a precursor to the RNA world in the PAH world hypothesis.

Multiple Genesis

Different forms of life may have appeared quasi-simultaneously in the early history of Earth. The other forms may be extinct, leaving distinctive fossils through their different biochemistry (e.g., using arsenic instead of phosphorus), survive as extremophiles, or simply be unnoticed through their being analoguous to organisms of the current life tree.

Criticisms

The modern concept of abiogenesis has been criticized by scientists throughout the years. Astronomer Sir Fred Hoyle did so based on the probability of abiogenesis randomly occurring. Physicist Hubert Yockey did so by saying that it is closer to theology than science.

Other scientists have proposed counterpoints to abiogenesis, such as, Harold Urey, Stanley Miller, Francis Crick (a molecular biologist), and Leslie Orgel's Directed Panspermia hypothesis.

Beyond making the trivial observation that life exists, it is difficult to prove or falsify abiogenesis; therefore the hypothesis has many such critics, both in the scientific and non-scientific communities. Nonetheless, research and hypothesizing continue in the hope of developing a satisfactory theoretical mechanism of abiogenesis.

Hoyle

Sir Fred Hoyle, with Chandra Wickramasinghe, was a critic of abiogenesis. Specifically Hoyle rejected chemical evolution to explain the naturalistic origin of life. His argument was mainly based on the improbability of what were thought to be the necessary components coming together for chemical evolution. Though modern theories address his argument, Hoyle never saw chemical evolution as a reasonable explanation. Hoyle preferred panspermia as an alternative natural explanation to the origin of life on Earth.

Yockey

Information theorist Hubert Yockey argued that chemical evolutionary research faces the following problem:

> Research on the origin of life seems to be unique in that the conclusion has already been authoritatively accepted ... . What remains to be done is to find the scenarios which describe the detailed mechanisms and processes by which this happened. One must conclude that, contrary to the established and current wisdom a scenario describing the genesis of life on earth by chance and natural causes which can be accepted on the basis of fact and not faith has not yet been written.

In a book he wrote 15 years later, Yockey argued that the idea of abiogenesis from a primordial soup is a failed paradigm:

> Although at the beginning the paradigm was worth consideration, now the entire effort in the primeval soup paradigm is self-deception on the ideology of its champions. ... The history of science shows that a paradigm, once it has achieved the status of acceptance (and is incorporated in textbooks) and regardless of its failures, is declared invalid only when a new paradigm is available to replace it. Nevertheless, in order to make progress in science, it is necessary to clear the

decks, so to speak, of failed paradigms. This must be done even if this leaves the decks entirely clear and no paradigms survive. It is a characteristic of the true believer in religion, philosophy and ideology that he must have a set of beliefs, come what may (Hoffer, 1951). Belief in a primeval soup on the grounds that no other paradigm is available is an example of the logical fallacy of the false alternative. In science it is a virtue to acknowledge ignorance. This has been universally the case in the history of science as Kuhn (1970) has discussed in detail. There is no reason that this should be different in the research on the origin of life. Yockey, in general, possesses a highly critical attitude toward people who give credence toward natural origins of life, often invoking words like "faith" and "ideology". Yockey's publications have become favorites to quote among creationists, though he is not a creationist himself (as noted in this 1995 email).

Abiogenic Synthesis of Key Chemicals

A number of problems with the RNA world hypothesis remain. There are no known chemical pathways for the abiogenic synthesis of nucleotides from pyrimidine nucleobases cytosine and uracil under prebiotic conditions. Other problems are the difficulty of nucleoside synthesis (from ribose and nucleobase), ligating nucleosides with phosphate to form the RNA backbone, and the short lifetime of the nucleoside molecules, especially cytosine which is prone to hydrolysis. Recent experiments also suggest that the original estimates of the size of an RNA molecule capable of self-replication were most probably vast underestimates. More-modern forms of the RNA World theory propose that a simpler molecule was capable of self-replication (that other "World" then evolved over time to produce the RNA World). At this time however, the various hypotheses have incomplete evidence supporting them. Many of them can be simulated and tested in the lab, but a lack of undisturbed sedimentary rock from that early in Earth's history leaves few opportunities to test this hypothesis robustly.

Homochirality Problem

Another unsolved issue in chemical evolution is the origin of homochirality, i.e. all monomers having the same "handedness" (amino acids being left handed, and nucleic acid sugars (ribose and deoxyribose) being right handed). Chiral molecules exist in nature as homogenous mixtures balanced approx. 50/50. This is called a racemic mixture. However, homochirality is essential for the formation of functional ribozymes and proteins. Proper formation is impeded by the very presence of right-handed amino acids and/or left-handed sugars in that they create malformed structures.

Clark has suggested that homochirality may have started in space, as the studies of the Amino acids on the Murchison meteorite showed L-analine to be more than twice as frequent as its D form, and L glutamic acid was more than 3 times its alternative counterpart. It is suggested that magnetically polarised light has the power to destroy one enantiomer within the proto-planetary disk. Once established chirality would be selected for.

Work performed in 2003 by scientists at Purdue identified the amino acid serine as being a probable root cause of the organic molecules' homochirality. Serine forms particularly strong bonds with amino acids of the same chirality, resulting in a cluster of eight molecules that must be all right-handed or left-handed. This property stands in contrast with other amino acids which are able to form weak bonds with amino acids of opposite chirality. Although the mystery of why left-handed serine became dominant is still unsolved, this result suggests an answer to the question of chiral transmission: how organic molecules of one chirality maintain dominance once asymmetry is established.

Relevant Fields

- Astrobiology is a field that may shed light on the nature of life in general, instead of just life as we know it on Earth, and may give clues as to how life originates.
- Complex systems

HISTORY OF BIOLOGY (BIOSCIENCE)

The history of biology (bioscience) traces the study of the living world from ancient to modern times. Although the concept of *biology* as a single coherent field arose in the 19th century, the biological sciences emerged from traditions of medicine and natural history reaching back to Galen and Aristotle in ancient Greece. During the Renaissance and early modern period, biological thought was revolutionized by a renewed interest in empiricism and the discovery of many novel organisms. Prominent in this movement were Vesalius and Harvey, who used experimentation and careful observation in physiology, and naturalists such as Linnaeus and Buffon who began to classify the diversity of life and the fossil record, as well as the development and behavior of organisms. Microscopy revealed the previously unknown world of microorganisms, laying the groundwork for cell theory. The growing importance of natural theology, partly a response to the rise of mechanical philosophy, encouraged the growth of natural history (though it entrenched the argument from design).

Over the 18th and 19th centuries, biological sciences such as botany and zoology became increasingly professional scientific disciplines. Lavoisier and other physical scientists began to connect the animate and inanimate worlds through physics and chemistry. Explorer-naturalists such as Alexander von Humboldt investigated the interaction between organisms and their environment, and the ways this relationship depends on geography—laying the foundations for biogeography, ecology and ethology. Naturalists began to reject essentialism and consider the importance of extinction and the mutability of species. Cell theory provided a new perspective on the fundamental basis of life. These developments, as well as the results from embryology and paleontology, were synthesized in Charles Darwin's theory of evolution by natural selection. The end of the 19th century saw the fall of spontaneous generation and the rise of the germ theory of disease, though the mechanism of inheritance remained a mystery.

In the early 20th century, the rediscovery of Mendel's work led to the rapid development of genetics by Thomas Hunt Morgan and his students, and by the 1930s the combination of population genetics and natural selection in the "neo-Darwinian synthesis". New disciplines developed rapidly, especially after Watson and Crick proposed the structure of DNA. Following the establishment of the Central Dogma and the cracking of the genetic code, biology was largely split between *organismal biology*—the fields that deal with whole organisms and groups of organisms—and the fields related to *cellular and molecular biology*. By the late 20th century, new fields like genomics and proteomics were reversing this trend, with organismal biologists using molecular techniques, and molecular and cell biologists investigating the interplay between genes and the environment, as well as the genetics of natural populations of organisms.

Etymology

The word *biology* is formed by combining the Greek âßïò (bios), meaning "life", and the suffix '-logy', meaning "science of", "knowledge of", "study of", based on the Greek verb ëÝãåéí, 'legein' = "to select", "to gather" (cf. the noun ëüãïò, 'logos' = "word"). The term *biology* in its modern sense appears to have been introduced independently by Karl Friedrich Burdach (in 1800), Gottfried Reinhold Treviranus (*Biologie oder Philosophie der lebenden Natur*, 1802) and Jean-Baptiste Lamarck (*Hydrogéologie*, 1802). The word itself appears in the title of Volume 3 of Michael Christoph Hanov's *Philosophiae naturalis sive physicae dogmaticae: Geologia, biologia, phytologia generalis et dendrologia*, published in 1766.

Before *biology*, there were several terms used for the study of animals and plants. *Natural history* referred to the descriptive aspects of biology, though it also included mineralogy and other non-biological fields; from the Middle Ages through the Renaissance, the unifying framework of natural history was the *scala naturae* or Great Chain of Being. *Natural philosophy* and *natural theology* encompassed the conceptual and metaphysical basis of plant and animal life, dealing with problems of why organisms exist and behave the way they do, though these subjects also included what is now geology, physics, chemistry, and astronomy. Physiology and (botanical) pharmacology were the province of medicine. *Botany*, *zoology*, and (in the case of fossils) *geology* replaced *natural history* and *natural philosophy* in the 18th and 19th centuries before *biology* was widely adopted.

Ancient and Medieval Knowledge

The earliest humans must have had and passed on knowledge about plants and animals to increase their chances of survival. This may have included knowledge of human and animal anatomy and aspects of animal behavior (such as migration patterns). However, the first major turning point in biological knowledge came with the Neolithic Revolution about 10,000 years ago. Humans first domesticated plants for farming, then livestock animals to accompany the resulting sedentary societies. The ancient cultures of Mesopotamia, Egypt, the Indian subcontinent, and China (among others) had sophisticated systems of philosophical, religious, and technical knowledge that encompassed the living world, and creation myths often centered on some aspect of life. However, the roots of modern biology are usually traced back to the secular tradition of ancient Greek philosophy. Frontispiece to a 1644 version of the expanded and illustrated edition of *Historia Plantarum* (ca. 1200), which was originally written around 200 BC

Ancient Greek

The Pre-Socratic philosophers asked many questions about life but produced little systematic knowledge of specifically biological interest—though the attempts of the atomists to explain life in purely physical terms would recur periodically through the history of biology. However, the medical theories of Hippocrates and his followers, especially humorism, had a lasting impact.

The philosopher Aristotle was the most influential scholar of the living world from antiquity. Though his early work in natural philosophy was speculative, Aristotle's later biological writings were more empirical, focusing on biological causation and the diversity of life. He made countless observations of nature, especially the habits and attributes of plants and animals in the world around him, which he devoted considerable attention to categorizing. In all, Aristotle classified 540 animal species, and dissected at least 50. He believed that intellectual purposes, formal causes, guided all natural processes. Aristotle, and nearly all scholars after him until the 18th century, believed that creatures were arranged in a graded scale of perfection rising from plants on up to humans: the *scala naturae* or Great Chain of Being. Aristotle's successor at the Lyceum, Theophrastus, wrote a series of books on botany—the *History of Plants*—which survived as the most important contribution of antiquity to botany, even into the Middle Ages. Many of Theophrastus' names survive into modern times, such as *carpos* for fruit, and *pericarpion* for seed vessel.

Pliny the Elder was also known for his knowledge of plants and nature, and was the most prolific compiler of zoological descriptions.

A few scholars in the Hellenistic period under the Ptolemies—particularly Herophilus of Chalcedon and Erasistratus of Chios—amended Aristotle's physiological work, even performing experimental dissections and vivisections. Claudius Galen became the most important authority on medicine and anatomy. Though a few ancient atomists such as Lucretius challenged the teleological Aristotelian viewpoint that all aspects of life are the result of design or purpose, teleology (and after the rise of Christianity, natural theology) would remain central to biological thought essentially until the 18th and 19th centuries. In the words of Ernst Mayr, "Nothing of any real consequence happened in biology after Lucretius and Galen until the Renaissance." The ideas of the Greek traditions of natural history and medicine survived, but they were generally taken unquestioningly.

Medieval Knowledge

De arte venandi, by Frederick II, Holy Roman Emperor, was an influential medieval natural history text that explored bird morphology. The decline of the Roman Empire led to the disappearance or destruction of much knowledge, though physicians still incorporated many aspects of the Greek tradition into training and practice. In Byzantium and the Islamic world, many of the Greek works were translated into Arabic and many of the works of Aristotle were preserved. During the High Middle Ages, a few European scholars such as Hildegard of Bingen, Albertus Magnus, and Frederick II expanded the natural history canon. The rise of European universities, though important for the development of physics and philosophy, had little impact on biological scholarship.

Renaissance and Early Modern Developments

The European Renaissance brought expanded interest in both empirical natural history and physiology. In 1543, Andreas Vesalius inaugurated the modern era of Western medicine with his seminal human anatomy treatise *De humani corporis fabrica*, which was based on dissection of corpses. Vesalius was the first in a series of anatomists who gradually replaced scholasticism with empiricism in physiology and medicine, relying on first-hand experience rather than authority and abstract reasoning. Via herbalism, medicine was also indirectly the source of renewed empiricism in the study of plants. Otto Brunfels, Hieronymus Bock and Leonhart Fuchs wrote extensively on wild plants, the beginning of a nature-based approach to the full range of plant life.

Bestiaries—a genre that combines both the natural and figurative knowledge of animals—also became more sophisticated, especially with the work of William

Turner, Pierre Belon, Guillaume Rondelet, Conrad Gessner, and Ulisse Aldrovandi. Artists such as Albrecht Dürer and Leonardo da Vinci, often working with naturalists, were also interested in the bodies of animals and humans, studying physiology in detail and contributing to the growth of anatomical knowledge. The traditions of alchemy and natural magic, especially in the work of Paracelsus, also laid claim to knowledge of the living world.

Alchemists subjected organic matter to chemical analysis and experimented liberally with both biological and mineral pharmacology. This was part of a larger transition in world views (the rise of the mechanical philosophy) that continued into the 17th century, as the traditional metaphor of *nature as organism* was replaced by the *nature as machine* metaphor.

17th and 18th Centuries

Cabinets of curiosities, such as that of Ole Worm, were centers of biological knowledge in the early modern period, bringing organisms from across the world together in one place. Before the Age of Exploration, naturalists had little idea of the sheer scale of biological diversity. Extending the work of Vesalius into experiments on still living bodies (of both humans and animals), William Harvey and other natural philosophers investigated the roles of blood, veins and arteries. Harvey's *De motu cordis* in 1628 was the beginning of the end for Galenic theory, and alongside Santorio Santorio's studies of metabolism, it served as an influential model of quantitative approaches to physiology.

In the early 17th century, the micro-world of biology was just beginning to open up. A few lensmakers and natural philosophers had creating crude microscopes since the late 16th century, and Robert Hooke published the seminal *Micrographia* based on observations with his own compound microscope in 1665. But it was not until Antony van Leeuwenhoek's dramatic improvements in lensmaking beginning in the 1670s—ultimately producing up to 200-fold magnification with a single lens—that scholars discovered spermatozoa, bacteria, infusoria and the sheer strangeness and diversity of microscopic life. Similar investigations by Jan Swammerdam led to new interest in entomology and built the basic techniques of microscopic dissection and staining.

In *Micrographia*, Robert Hooke had applied the word *cell* to biological structures such as this piece of cork, but it was not until the 19th century that scientists considered cells the universal basis of life.

As the microscopic world was expanding, the macroscopic world was shrinking. Botanists such as John Ray worked to incorporate the flood of newly discovered organisms shipped from across the globe into a coherent taxonomy, and a coherent theology (natural theology). Debate over another flood, the Noachian, catalyzed the development of paleontology; in 1669 Nicholas Steno published an essay on how the remains of living organisms could be trapped in layers of sediment and mineralized to produce fossils. Although Steno's ideas about fossilization were well known and much debated among natural philosophers, an organic origin for all fossils would not be accepted by all naturalists until the end of the 18th century due to philosophical and theological debate about issues such as the age of the earth and extinction. Systematizing, naming and classifying dominated natural history throughout much of the 17th and 18th centuries. Carolus Linnaeus published a basic taxonomy for the natural world in 1735 (variations of which have been in use ever since), and in the 1750s introduced scientific names for all his species. While Linnaeus conceived of species as unchanging parts of a designed hierarchy, the other great naturalist of the 18th century, Georges-Louis Leclerc, Comte de Buffon, treated species as artificial categories and living forms as malleable—even suggesting the possibility of common descent. Though he was opposed to evolution, Buffon is a key figure in the history of evolutionary thought; his work would influence the evolutionary theories of both Lamarck and Darwin.

The discovery and description of new species and the collection of specimens became a passion of scientific gentlemen and a lucrative enterprise for entrepreneurs; many naturalists traveled the globe in search of scientific knowledge and adventure.

17th Century: The Emergence of Biological Disciplines

Up through the nineteenth century, the scope of biology was largely divided between medicine, which investigated questions of form and function (*i.e.*, physiology), and natural history, which was concerned with the diversity of life and interactions among different forms of life and between life and non-life. By 1900, much of these domains overlapped, while natural history and (and its counterpart natural philosophy) had largely given way to more specialized scientific disciplines—cytology, bacteriology, morphology, embryology, geography, and geology.

Cell theory and Germ Theory

Advances in microscopy also had a profound impact on biological thinking: in 1839, Schleiden and Schwann proposed the cell theory—that the basic unit of organisms is the cell and all cells come from preexisting cells. The cytologist Walther Flemming in 1882 was the first to demonstrate that the discrete stages of mitosis were not an artifact of staining, but occurred in living cells, and moreover, that chromosomes doubled in number just before the cell divided and a daughter cell was produced. In 1887 August Weismann proposed that the chromosome number must then be halved in the case of the sexual cells, the gametes. This was shortly proved to be the case and the process of meiosis began to be understood.

By the mid 1850s the miasma theory of disease was largely superseded by the germ theory of disease, creating extensive interest in microorganisms and their interactions with other forms of life. By the 1880s, bacteriology was becoming a coherent discipline, especially through the work of Robert Koch, who introduced methods for growing pure cultures on agar gels containing specific nutrients in Petri dishes. The long-held idea that living organisms could easily originate from nonliving matter (spontaneous generation) was attacked in a series of experiments carried out by Louis Pasteur, while debates over vitalism vs. mechanism (a perennial issue since the time of Aristotle and the Greek atomists) continued apace.

Physiology

Over the course of the 19th century, the scope of physiology expanded greatly, from a primarily medically-oriented field to a wide-ranging investigation of the physical and chemical processes of life—including plants, animals, and even microorganisms in addition to man. *Living things as machines* became a dominant metaphor in biological (and social) thinking. Innovative laboratory glassware and experimental methods developed by Louis Pasteur and other biologists contributed to the young field of bacteriology in the late 19th century.

Rise of Experimental Physiology and Organic Chemistry

In chemistry, one central issue was the distinction between organic and inorganic substances, especially in the context of organic transformations such as fermentation and putrefaction. Since Aristotle these had been considered essentially biological (*vital*) processes. However, Friedrich Wöhler, Justus Liebig and other pioneers of the rising field of organic chemistry—building on the work of Lavoisier—showed that the organic world could often be analyzed by physical and chemical methods. In 1828 Wöhler showed that the organic substance urea could be created by chemical means that do not involve life, providing a powerful argument against vitalism.

Cell extracts ("ferments") that could effect chemical transformations were discovered, beginning with diastase in 1833, and by the end of the 19th century the concept of enzymes was well established, though equations of chemical kinetics would not be applied to enzymatic reactions until the early 20th century. Physiologists such as Claude Bernard explored (through vivisection and other experimental methods) the chemical and physical functions of living bodies to an unprecedented degree, laying the groundwork for endocrinology (a field that developed quickly after the discovery of the first hormone, secretin, in 1902), biomechanics, and the study of nutrition and digestion.

The importance and diversity of experimental physiology methods, within both medicine and biology, grew dramatically over the second half of the 19th century.

The control and manipulation of life processes became a central concern, and experiment was placed at the center of biological education.

Natural Philosophy and Natural History

In the course of his travels, Alexander von Humboldt mapped the distribution of plants across landscapes and recorded a variety of physical conditions such as pressure and temperature. Widespread travel by naturalists in the early- to mid-nineteenth century resulted in a wealth of new information about the diversity and distribution of living organisms. Of particular importance was the work of Alexander von Humboldt, which analyzed the relationship between organisms and their environment (*i.e.*, the domain of natural history) using the quantitative approaches of natural philosophy (*i.e.*, physics and chemistry). Humboldt's work laid the foundations of biogeography and inspired several generations of scientists.

Evolution and Biogeography

The first to propose an evolutionary theory was Jean-Baptiste Lamarck; based on the inheritance of acquired characteristics (an inheritance mechanism that was widely accepted until the 20th century), it described a chain of development stretching from the lowliest microbe to humans. The British naturalist Charles Darwin, combining the biogeographical approach of Humboldt, the uniformitarian geology of Lyell, Thomas Malthus's writings on population growth, and his own morphological expertise, created a more successful evolutionary theory based on natural selection; similar evidence lead Alfred Russel Wallace to independently reach the same conclusions. The 1859 publication of Darwin's theory in *On the Origin of Species by Means of Natural Selection, or the Preservation of Favoured Races in the Struggle for Life* is often considered the central event in the history of modern biology. Darwin's established credibility as a naturalist, the sober tone of the work, and most of all the sheer strength and volume of evidence presented, allowed *Origin* to succeed where previous evolutionary works such as the anonymous *Vestiges of Creation* had failed.

Most scientists were convinced of evolution and common descent by the end of the 19th century. However, natural selection would not be accepted as the primary mechanism of evolution until well into the 20th century, as most contemporary theories of heredity seemed incompatible with the inheritance of random variation. Wallace, building on earlier work by Humbolt and Darwin, made major contributions to biogeography by focusing on the distribution of closely allied species with particular attention to the effects of geographical barriers during his research in the Amazon basin and the Malay archipelago. He discovered the Wallace line dividing the fauna of the Malay archipelago between a zone allied with Asia and a zone allied with Australia.

The ornithologist Philip Sclater, drawing on the work of Wallace and others, proposed a system of 6 major geographical regions to describe the distribution of bird species in the world. Wallace and others would in turn extend Sclater's system from birds to animals of all kinds. The scientific study of heredity grew rapidly in the wake of Darwin's *On the Origin of Species* (1859) with the work of Francis Galton and the biometricians.

The origin of genetics is usually traced to the 1866 work of the monk Gregor Mendel, who would later be credited with the laws of inheritance. However, his work was not recognized as significant until 35 years afterward. In the meantime, a variety of theories of inheritance (based on pangenesis, orthogenesis, or other mechanisms) were debated and investigated vigorously. Embryology and ecology also became central biological fields, especially as linked to evolution and popularized in the work of Ernst Haeckel.

Geology and Paleontology

The emerging discipline of geology also brought natural history and natural philosophy closer together; the establishment of the stratigraphic column linked the spacial distribution of organisms to their temporal distribution, a key precursor to concepts of evolution. Georges Cuvier and others made great strides in comparative anatomy and paleontology in the late 1790s and early 1800s. In a series of lectures and papers that made detailed comparisons between living mammals and fossil remains Cuvier was able to establish that the fossils were remains of species that had become extinct—rather than being remains of species still alive elsewhere in the world, as had been widely believed. Fossils discovered and described by Gideon Mantell, William Buckland, Mary Anning, and Richard Owen among others helped establish that there had been an 'age of reptiles' that had preceded even the prehistoric mammals. These discoveries captured the public imagination and focused attention on the history of life on earth. Most of these geologists held to catastrophism, but Charles Lyell's influential *Principles of Geology* (1830) popularised Hutton's uniformitarianism, a theory that explained the geological past and present on equal terms. Charles Darwin's first sketch of an evolutionary tree from his *First Notebook on Transmutation of Species* (1837)

20th Century Biological Sciences

At the beginning of the 20th century, biological research was largely a professional endeavour. However, most work was still done in the natural history mode, which emphasized morhphological and phylogenetic analysis over experiment-based causal explanations. However, anti-vitalist experimental physiologists and embryologists, especially in Europe, were increasingly influential. The tremendous success of experimental approaches to development, heredity, and metabolism in the 1900s and 1910s demonstrated the power of experimentation in biology. In the following decades, experimental work replaced natural history as the dominant mode of research.

Ecology and Environmental Science

In the early 20th century, naturalists were faced with increasing pressure to add rigor and preferably experimentation to their methods, as the newly prominent laboratory-based biological disciplines had done. Ecology had emerged as a combination of biogeography with the biogeochemical cycle concept pioneered by chemists; field biologists developed quantitative methods such as the quadrat and adapted laboratory instruments and cameras for the field to further set their work apart from traditional natural history. Zoologists and botanists did what they could to mitigate the unpredictability of the living world, performing laboratory experiments and studying semi-controlled natural environments such as gardens; new institutions like the Carnegie Station for Experimental Evolution and the Marine Biological Laboratory provided more controlled environments for studying organisms through their entire life cycles.

The ecological succession concept, pioneered in the 1900s and 1910s by Henry Chandler Cowles and Frederic Clements, was important in early plant ecology. Alfred Lotka's predator-prey equations, G. Evelyn Hutchinson's studies of the biogeography and biogeochemical structure of lakes and rivers (limnology) and Charles Elton's studies of animal food chains were pioneers among the succession of quantitative methods that colonized the developing ecological specialties. Ecology became an independent discipline in the 1940s and 1950s after Eugene P. Odum synthesized many of the concepts of ecosystem ecology, placing relationships between groups of organisms (especially material and energy relationships) at the center of the field.

In the 1960s, as evolutionary theorists explored the possibility of multiple units of selection, ecologists turned to evolutionary approaches. In population ecology, debate over group selection was brief but vigorous; by 1970, most biologists agreed that natural selection was rarely effective above the level of individual organisms. The evolution of ecosystems, however, became a lasting research focus. Ecology expanded rapidly with the rise of the environmental movement; the International Biological Program attempted to apply the methods of big science (which had been so successful in the physical sciences) to ecosystem ecology and pressing environmental issues, while smaller-scale independent efforts such as island biogeography and the Hubbard Brook Experimental Forest helped redefine the scope of an increasingly diverse discipline.

Biochemistry, Microbiology, and Molecular Biology

By the end of the 19th century all of the major pathways of drug metabolism had been discovered, along with the outlines of protein and fatty acid metabolism and urea synthesis. In the early decades of the twentieth century, the minor components of foods in human nutrition, the vitamins, began to be isolated and synthesized. Improved laboratory techniques such as chromatography and electrophoresis led to rapid advances in physiological chemistry, which—as *biochemistry*—began to achieve independence from its medical origins. In the 1920s and 1930s, biochemists—led by Hans Krebs and Carl and Gerty Cori—began to work out many of the central metabolic pathways of life: the citric acid cycle, glycogenesis and glycolysis, and the synthesis of steroids and porphyrins. Between the 1930s and 1950s, Fritz Lipmann and others established the role of ATP as the universal carrier of energy in the cell, and mitochondria as the powerhouse of the cell. Such traditionally biochemical work continued to be very actively pursued throughout the 20th century and into the 21st.

Classical Genetics, the Modern Synthesis, and Evolutionary Theory

1900 marked the so-called *rediscovery of Mendel*: Hugo de Vries, Carl Correns, and Erich von Tschermak independently arrived at Mendel's laws (which were not actually present in Mendel's work). Soon after, cytologists (cell biologists) proposed that chromosomes were the hereditary material. Between 1910 and 1915, Thomas Hunt Morgan and the "Drosophilists" in his fly lab forged these two ideas—both controversial—into the "Mendelian-chromosome theory" of heredity. They quantified the phenomenon of genetic linkage and postulated that genes reside on chromosomes like beads on string; they hypothesized crossing over to explain linkage and constructed genetic maps of the fruit fly *Drosophila melanogaster*, which became a widely used model organism.

Hugo de Vries tried to link link the new genetics with evolution; building on his work with heredity and hybridization, he proposed a theory of mutationism, which was widely accepted in the early 20th century. Lamarckism also had many adherents. Darwinism was seen as incompatible with the continuously variable traits studied by biometricians, which seemed only partially heritable. In the 1920s and 1930s—following the acceptance of the Mendelian-chromosome theory— the emergence of the discipline of population genetics, with the work of R.A. Fisher, J.B.S. Haldane and Sewall Wright, unified the idea of evolution by natural selection with Mendelian genetics producing the modern synthesis. The inheritance of acquired characters was rejected, while mutationism gave way as genetic theories matured.

In the second half of the century the ideas of population genetics began to be applied in the new discipline of the genetics of behavior, sociobiology, and, especially in humans, evolutionary psychology. In the 1960s W.D. Hamilton and others developed game theory approaches to explain altruism from an evolutionary perspective through kin selection. The possible origin of higher organisms through endosymbiosis, and contrasting approaches to molecular evolution in the gene-centered view (which held selection as the predominant cause of evolution) and the neutral theory (which made genetic drift a key factor) spawned perennial debates over the proper balance of adaptationism and contingency in evolutionary theory.

In the 1970s Stephen Jay Gould and Niles Eldredge proposed the theory of punctuated equilibrium which holds that stasis is the most prominent feature of the fossil record, and that most evolutionary changes occur rapidly over relatively short periods of time. In 1980 Luis Alvarez and Walter Alvarez proposed the hypothesis that an impact event was responsible for the Cretaceous-Tertiary extinction event. Also in the early 1980s, statistical analysis of the fossil record of marine organisms published by Jack Sepkoski and David M. Raup lead to a better appreciation of the importance of mass extinction events to the history of life on earth.

Molecular Biology Origins

Following the rise of classical genetics, many biologists—including a new wave of physical scientists in biology—pursued the question of the gene and its physical nature. Warren Weaver—head of the science division of the Rockefeller Foundation—issued grants to promote research that applied the methods of physics and chemistry to basic biological problems, coining the term *molecular biology* for this approach in 1938; many of the significant biological breakthroughs of the 1930s and 1940s were funded by the Rockefeller Foundation. Wendell Stanley's crystallization of tobacco mosaic virus as a pure nucleoprotein in 1935 convinced many scientists that heredity might be explained purely through physics and chemistry.

Like biochemistry, the overlapping disciplines of bacteriology and virology (later combined as *microbiology*), situated between science and medicine, developed rapidly in the early 20th century. Félix d'Herelle's isolation of bacteriophage during World War I initiated a long line of research focused of phage viruses and the bacteria they infect. The development of standard, genetically uniform organisms that could produce repeatable experimental results was essential for the development of molecular genetics. After early work with *Drosophila* and maize, the adoption of simpler model systems like the bread mold *Neurospora crassa* made it possible to connect genetics to biochemistry, most importantly with Beadle and Tatum's "one gene, one enzyme" hypothesis in 1941.

Genetics experiments on even simpler systems like tobacco mosaic virus and bacteriophage, aided by the new technologies of electron microscopy and ultracentrifugation, forced scientists to re-evaluate the literal meaning of *life*; virus heredity and reproducing nucleoprotein cell structures outside the nucleus ("plasmagenes") complicated the accepted Mendelian-chromosome theory.

The "central dogma of molecular biology" (originally a "dogma" only in jest) was proposed by Francis Crick in 1958. This is Crick's reconstruction of how he conceived of the central dogma at the time. The solid lines represent (as it seemed in 1958) known modes of information transfer, and the dashed lines represent postulated ones. Oswald Avery showed in 1943 that DNA was likely the genetic material of the chromosome, not its protein; the issue was settled decisively with the 1952 Hershey-Chase experiment—one of many contribution from the so-called phage group centered around physicist-turned-biologist Max Delbrück. In 1953 James D. Watson and Francis Crick, building on the work of Maurice Wilkins and Rosalind Franklin, suggested that the structure of DNA was a double helix.

In their famous paper "Molecular structure of Nucleic Acids", Watson and Crick noted coyly, "It has not escaped our notice that the specific pairing we have postulated immediately suggests a possible copying mechanism for the genetic material." After the 1958 Meselson-Stahl experiment confirmed the semiconservative replication of DNA, it was clear to most biologists that nucleic acid sequence must somehow determine amino acid sequence in proteins; physicist George Gamow proposed that a fixed genetic code connected proteins and DNA. Between 1953 and 1961, there were few known biological sequences—either DNA or protein—but an abundance of proposed code systems, a situation made even more complicated by expanding knowledge of the intermediate role of RNA. To actually decipher the code, it took an extensive series of experiments in biochemistry and bacterial genetics, between 1961 and 1966—most importantly the work of Nirenberg and Khorana.

Expansion of Molecular Biology

In addition to the Division of Biology at Caltech, the Laboratory of Molecular Biology (and its precursors) at Cambridge, and a handful of other institutions, the Pasteur Institute became a major center for molecular biology research in the late 1950s. Scientists at Cambridge, led by Max Perutz and John Kendrew, focused on the rapidly developing field of structural biology, combining X-ray crystallography with molecular modelling and the new computational possibilities of digital computing (benefiting both directly and indirectly from the military funding of science). A number of biochemists led by Fred Sanger later joined the Cambridge lab, bringing together the study of macromolecular structure and function. At the Pasteur Institute, François Jacob and Jacques Monod followed the 1959 PaJaMo experiment with a series of publications regarding the *lac* operon that established the concept of gene regulation and identified what came to be known as messenger RNA. By the mid-1960s, the intellectual core of molecular biology—a model for the molecular basis of metabolism and reproduction— was largely complete.

The late 1950s to the early 1970s was a period of intense research and institutional expansion for molecular biology, which had only recently become a somewhat coherent discipline. In what organismic biologist E. O. Wilson called "The Molecular Wars", the methods and practitioners of molecular biology spread rapidly, often coming to dominate departments and even entire disciplines. Molecularization was particularly important in genetics, immunology, embryology, and neurobiology, while the idea that life is controlled by a "genetic program"—a metaphor Jacob and Monod introduced from the emerging fields of cybernetics and computer science—became an influential perspective throughout biology. Immunology in particular became linked with molecular biology, with innovation flowing both ways: the clonal selection theory developed by Niels Jerne and Frank Macfarlane Burnet in the mid 1950s helped shed light on the general mechanisms of protein synthesis. Resistance to the growing influence molecular biology was especially evident in evolutionary biology. Protein sequencing had great potential for the quantitative study of evolution (through the molecular clock hypothesis), but leading evolutionary biologists questioned the relevance of molecular biology for answering the big questions of evolutionary causation.

Departments and disciplines fractured as organismic biologists asserted their importance and independence: Theodosius Dobzhansky made the famous statement that "nothing in biology makes sense except in the light of evolution" as a response to the molecular challenge. The issue became even more critical after 1968; Motoo Kimura's neutral theory of molecular evolution suggested natural selection was not the ubiquitous cause of evolution, at least at the molecular level, and that molecular evolution might be a fundamentally different process from morphological evolution. (Resolving this "molecular/morphological paradox" has been a central focus of molecular evolution research since the 1960s.)

Biotechnology, Genetic Engineering, and Genomics

Biotechnology in the general sense has been an important part of biology since the late 19th century. With the industrialization of brewing and agriculture, chemists and biologists became aware of the great potential of human-controlled biological processes. In particular, fermentation proved a great boon to chemical industries. By the early 1970s, a wide range of biotechnologies were being developed, from drugs like penicillin and steroids to foods like *Chlorella* and single-cell protein to gasohol—as well as a wide range of hybrid high-yield crops and agricultural technologies, the basis for the Green Revolution.

Biotechnology in the modern sense of genetic engineering began in the 1970s, with the invention of recombinant DNA techniques. Restriction enzymes were discovered and characterized in the late 1960s, following on the heels of the isolation, then duplication, then synthesis of viral genes. Beginning with the lab of Paul Berg in 1972 (aided by *EcoRI* from Herbert Boyer's lab, building on work with ligase by Arthur Kornberg's lab), molecular biologists put these pieces together to produce the first transgenic organisms. Soon after, others began using plasmid vectors and adding genes for antibiotic resistance, greatly increasing the reach of the recombinant techniques. Wary of the potential dangers (particularly the possibility of a prolific bacteria with a viral cancer-causing gene), the scientific community as well as a wide range of scientific outsiders reacted to these developments with both enthusiasm and fearful restraint. Prominent molecular biologists led by Berg suggested a temporary moratorium on recombinant DNA research until the dangers could be assessed and policies could be created.

This moratorium was largely respected, until the participants in the 1975 Asilomar Conference on Recombinant DNA created policy recommendations and concluded that the technology could be used safely. Following Asilomar, new genetic engineering techniques and applications developed rapidly. DNA sequencing methods improved greatly (pioneered by Fred Sanger and Walter Gilbert), as did oligonucleotide synthesis and transfection techniques.

Researchers learned to control the expression of transgenes, and were soon racing—in both academic and industrial contexts—to create organisms capable of expressing human genes for the production of human hormones. However, this was a more daunting task than molecular biologists had expected; developments between 1977 and 1980 showed that, due to the phenomena of split genes and splicing, higher organisms had a much more complex system of gene expression than the bacteria models of earlier studies. The first such race, for synthesizing human insulin, was won by Genentech. This marked the beginning of the biotech boom (and with it, the era of gene patents), with an unprecedented level of overlap between biology, industry, and law.

Molecular Systematics and Genomics

Inside of a 48-well thermal cycler, a device used to perform polymerase chain reaction on many samples at once. By the 1980s, protein sequencing had already transformed methods of scientific classification of organisms (especially cladistics) but biologists soon began to use RNA and DNA sequences as characters; this expanded the significance of molecular evolution within evolutionary biology, as the results of molecular systematics could be compared with traditional evolutionary trees based on morphology. Following the pioneering ideas of Lynn Margulis on endosymbiotic theory, which holds that some of the organelles of eukaryotic cells originated from free living prokaryotic organisms through symbiotic relationships, even the overall division of the tree of life was revised. Into the 1990s, the five domains (Plants, Animals, Fungi, Protists, and Monerans) became three (the Archaea, the Bacteria, and the Eukarya) based on Carl Woese's pioneering molecular systematics work with 16S rRNA sequencing.

The development and popularization of the polymerase chain reaction (PCR) in mid 1980s (by Kary Mullis and others at Cetus Corp.) marked another watershed in the history of modern biotechnology, greatly increasing the ease and speed of genetic analysis.

Coupled with the use of expressed sequence tags, PCR led to the discovery of many more genes than could be found through traditional biochemical or genetic methods and opened the possibility of sequencing entire genomes. The unity of much of the morphogenesis of organisms from fertilized egg to adult began to be unraveled after the discovery of the homeobox genes, first in fruit flies, then in other insects and animals, including humans. These developments led to advances in the field of evolutionary developmental biology towards understanding how the various body plans of the animal phyla have evolved and how they are related to one another.

The Human Genome Project—the largest, most costly single biological study ever undertaken—began in 1988 under the leadership of James D. Watson, after preliminary work with genetically simpler model organisms such as *E. coli*, *S. cerevisiae* and *C. elegans*. Shotgun sequencing and gene discovery methods pioneered by Craig Venter—and fueled by the financial promise gene patents with Celera Genomics— led to a public-private sequencing competition that ended in compromise with the first draft of the human DNA sequence announced in 2000.

TIMELINE OF BIOLOGY AND ORGANIC CHEMISTRY

A timeline of significant events in biology and organic chemistry

Before 1600

c. 300 B.C. - Alcmaeon of Croton distinguished veins from arteries and discovered the optic nerve.

c. 200 B.C. - Sushruta - wrote Sushruta Samhita describing over 120 surgical instruments, 300 surgical procedures and classified human surgery in 8 categories. Performed cosmetic surgery.

c. 400 B.C. - Xenophanes examined fossils and speculated on the evolution of life.

c.700 B.C. - Aristotle attempted a comprehensive classification of animals. His written works include Historion Animalium, a general biology of animals, De Partibus

Animalium, a comparative anatomy and physiology of animals, and De Generatione Animalium, on developmental biology.

c. 600 BC - Theophrastos (or Theophrastus) begins the systematic study of botany.

c. 900 B.C. - Herophilos dissected the human body.

c. 100 B.C. - Diocles wrote the last known anatomy book and was the first to use the term anatomy.

c. 50-70 - Historia Naturalis by Pliny the Elder (Gaius Plinius Secundus) was published in 37 volumes.

130-200 - Claudius Galen wrote numerous treatises on human anatomy.

c. 1010 - Avicenna (Ibn Sina or Abu Ali al Hussein ibn Abdallah) published his Canon of Medicine (Kitab al-Qanun fi al-tibb).

1543 - Andreas Vesalius publishes the anatomy treatise De humani corporis fabrica.

1600-1699

?? - Jan Baptist van Helmont performs his famous tree plant experiment in which he shows that the substance of a plant derives from water and air, the first description of photosynthesis.

1628 - William Harvey publishes An Anatomical Exercise on the Motion of the Heart and Blood in Animals

1651 - William Harvey concludes that all animals, including mammals, develop from eggs, and spontaneous generation of any animal from mud or excrement was an impossibility.

1658 - Jan Swammerdam observes red blood cells under a microscope.

1663 - Robert Hooke sees cells in cork using a microscope.

1668 - Francesco Redi disproves spontaneous generation by showing that fly maggots only appear on pieces of meat in jars if the jars are open to the air. Jars covered with cheesecloth contained no flies.

1672 - Marcello Malpighi publishes the first description of chick development, including the formation of muscle somites, circulation, and nervous system.

1676 - Anton van Leeuwenhoek observes protozoa and calls them animalcules.

1677 - Anton van Leeuwenhoek observes spermatozoa.

1683 - Anton van Leeuwenhoek observes bacteria. Leeuwenhoek's discoveries renew the question of spontaneous generation in microorganisms.

1700-1799

1767 - Kaspar Friedrich Wolff argues that the tissues of a developing chick form from nothing and are not simply elaborations of already-present structures in the egg.

1768 - Lazzaro Spallanzani again disproves spontaneous generation by showing that no organisms grow in a rich broth if it is first heated (to kill any organisms) and allowed to cool in a stoppered flask. He also shows that fertilization in mammals requires an egg and semen.

1771 - Joseph Priestley demonstrates that plants produce a gas that animals and flames consume. Those two gases are carbon dioxide and oxygen.

1798 - Thomas Malthus discusses human population growth and food production in An Essay on the Principle of Population.

1800-1899

1801 - Jean-Baptiste Lamarck begins the detailed study of invertebrate taxonomy.

1802 - The term biology in its modern sense is propounded independently by Gottfried Reinhold Treviranus (Biologie oder Philosophie der lebenden Natur) and Lamarck (Hydrogéologie). The word had been coined in 1800 by Karl Friedrich Burdach.

1809 - Lamarck proposes a modern theory of evolution based on the inheritance of acquired characteristics.

1817 - Pierre-Joseph Pelletier and Joseph-Bienaime Caventou isolate chlorophyll.

1820 - Christian Friedrich Nasse formulates Nasse's law: hemophilia occurs only in males and is passed on by unaffected females.

1824 - J.L Prevost and J.B. Dumas showed that the sperm in semen were not parasites, as previously thought, but, instead, the agents of fertilization.

1826 - Karl von Baer shows that the eggs of mammals are in the ovaries, ending a 200-year search for the mammalian egg.

1828 - Friedrich Woehler synthesizes urea; first synthesis of an organic compound from inorganic starting materials.i am cool

1836 - Theodor Schwann discovers pepsin in extracts from the stomach lining; first isolation of an animal enzyme.

1837 - Theodor Schwann shows that heating air will prevent it from causing putrefaction.

1838 - Matthias Schleiden proposes that all plants are composed of cells.

1839 - Theodor Schwann proposes that all animal tissues are composed of cells. Schwann and Schleinden argued that cells are the elementary particles of life.

1856 - Louis Pasteur states that microorganisms produce fermentation.

1858 - Charles R. Darwin and Alfred Wallace independently propose a theory of biological evolution ("descent through modification") by means of natural selection. Only in later editions of his works did Darwin used the term "evolution."

1858 - Rudolf Virchow proposes that cells can only arise from pre-existing cells; "Omnis cellula e celulla," all cell from cells. The Cell Theory states that all organisms are composed of cells (Schleiden and Schwann), and cells can only come from other cells (Virchow).

1864 - Louis Pasteur disproves the spontaneous generation of cellular life.

1865 - Gregor Mendel demonstrates in pea plants that inheritance follows definite rules. The Principle of Segregation states that each organism has two genes per trait, which segregate when the organism makes eggs or sperm. The Principle of Independent Assortment states that each gene in a pair is distributed independently during the formation of eggs or sperm. Mendel's trailblazing foundation for the science of genetics went unnoticed, to his lasting disappointment.

1865 - Friedrich August Kekulé von Stradonitz realizes that benzene is composed of carbon and hydrogen atoms in a hexagonal ring.

1869 - Friedrich Miescher discovers nucleic acids in the nuclei of cells.

1874 - Jacobus van 't Hoff and Joseph-Achille Le Bel advance a three-dimensional stereochemical representation of organic molecules and propose a tetrahedral carbon atom.

1876 - Oskar Hertwig and Hermann Fol independently describe (in sea urchin eggs) the entry of sperm into the egg and the subsequent fusion of the egg and sperm nuclei to form a single new nucleus.

1884 - Emil Fischer begins his detailed analysis of the compositions and structures of sugars.

1892 - Hans Driesch separates the individual cells of a 2-cell sea urchin embryo and shows that each cell develops into a complete individual, thus disproving the theory of preformation and showing that each cell is "totipotent," containing all the hereditary information necessary to form an individual.

1898 - Martinus Beijerinck uses filtering experiments to show that tobacco mosaic disease is caused by something smaller than a bacterium, which he names a virus.

1900-1949

1900 - Two biologists independently rediscover Mendel's paper on heredity.

1902 - Walter Sutton and Theodor Boveri, independently propose that the chromosomes carry the hereditary information.

1905 - William Bateson coins the term "genetics" to describe the study of biological inheritance.

1906 - Mikhail Tsvet discovers the chromatography technique for organic compound separation.

1907 - Ivan Pavlov demonstrates conditioned responses with salivating dogs.

1907 - Emil Fischer artificially synthesizes peptide amino acid chains and thereby shows that amino acids in proteins are connected by amino group-acid group bonds.

1909 - Wilhelm Johannsen coined the word "gene."

1911 - Thomas Hunt Morgan proposes that genes are arranged in a line on the chromosomes.

1926 - James Sumner shows that the urease enzyme is a protein.

1928 - Otto Diels and Kurt Alder discover the Diels-Alder cycloaddition reaction for forming ring molecules.

1928 - Alexander Fleming discovers the first antibiotic, penicillin

1929 - Phoebus Levene discovers the sugar deoxyribose in nucleic acids.

1929 - Edward Doisy and Adolf Butenandt independently discover estrone.

1930 - John Howard Northrop shows that the pepsin enzyme is a protein.

1931 - Adolf Butenandt discovers androsterone.

1932 - Hans Adolf Krebs discovers the urea cycle.

1933 - Tadeus Reichstein artificially synthesizes vitamin C; first vitamin synthesis.

1935 - Rudolf Schoenheimer uses deuterium as a tracer to examine the fat storage system of rats.

1935 - Wendell Stanley crystallizes the tobacco mosaic virus.

1935 - Konrad Lorenz describes the imprinting behavior of young birds.

1937 - Dorothy Crowfoot Hodgkin discovers the three-dimensional structure of cholesterol.

1937 - Hans Adolf Krebs discovers the tricarboxylic acid cycle.

1937 - In Genetics and the Origin of Species, Theodosius Dobzhansky applies the chromosome theory and population genetics to natural populations in the first mature work of neo-Darwinism, also called the modern synthesis, a term coined by Julian Huxley.

1938 - A living coelacanth is found off the coast of southern Africa.

1940 - Donald Griffin and Robert Galambos announce their discovery of sonar echolocation by bats.

1942 - Max Delbruck and Salvador Luria demonstrate that bacterial resistance to virus infection is caused by random mutation and not adaptive change.

1944 - Oswald Avery shows that DNA carries the genetic code in pneumococcus bacteria.

1944 - Robert Burns Woodward and William von Eggers Doering synthesize quinine.

1945 - Dorothy Crowfoot Hodgkin discovers the three-dimensional structure of penicillin.

1948 - Erwin Chargaff shows that in DNA the number of guanine units equals the number of cytosine units and the number of adenine units equals the number of thymine units.

1950-1989

1951 - Robert Woodward synthesizes cholesterol and cortisone.

1952 - American developmental biologists Robert Briggs and Thomas King clone the first vertebrate by transplanting nuclei from leopard frogs embryos into enucleated eggs. More differentiated cells were the less able they are to direct development in the enucleated egg.

1952 - Alfred Hershey and Martha Chase show that DNA is the genetic material in bacteriophage viruses.

1952 - Fred Sanger, Hans Tuppy, and Ted Thompson complete their chromatographic analysis of the insulin amino acid sequence.

1952 - Rosalind Franklin concludes that DNA is a double helix with a diameter of 2 nm and the sugar-phosphate backbones on the outside of the helix, based on x ray diffraction studies. She suspects the two sugar-phosphate backbones have a peculiar relationship to each other.

1953 - After examining Franklin's unpublished data, James D. Watson and Francis Crick publish a double-helix structure for DNA, with one sugar-phosphate backbone running in the opposite direction to the other. They further suggest a mechanism by which the molecule can replicate itself and serve to transmit genetic information. Their paper, combined with the Hershey-Chase experiment and Chargaff's data on nucleotides, finally persuades biologists that DNA is the genetic material, not protein.

1953 - Max Perutz and John Kendrew determine the structure of hemoglobin using X-ray diffraction studies.

1953 - Stanley Miller shows that amino acids can be formed when simulated lightning is passed through vessels containing water, methane, ammonia, and hydrogen

1954 - Dorothy Crowfoot Hodgkin discovers the three-dimensional structure of vitamin B-12.

1955 - Marianne Grunberg-Manago and Severo Ochoa discover the first nucleic-acid-synthesizing enzyme (polynucleotide phosphorylase), which links nucleotides together into polynucleotides.

1955 - Arthur Kornberg discovers DNA polymerase enzymes.

1958 - Matthew Stanley Meselson and Franklin W. Stahl prove that DNA replication is semiconservative in the Meselson-Stahl experiment

1959 - Severo Ochoa and Arthur Kornberg receive a Nobel Prize for their work.

1959 - Max Perutz describes the structure of hemoglobin, the oxygen-carrying protein in blood.

1960 - John Kendrew describes the structure of myoglobin, the oxygen-carrying protein in muscle.

1960 - Four separate researchers (S. Weiss, J. Hurwitz, Audrey Stevens and J. Bonner) discover bacterial RNA polymerase, which polymerizes nucleotides under the direction of DNA.

1960 - Juan Oro finds that concentrated solutions of ammonium cyanide in water can produce the nucleotide organic base adenine.

1960 - Robert Woodward synthesizes chlorophyll.

1961 - German plant physiologist H. J. Matthaei cracks the first codon of the genetic code (the codon for the amino acid phenylalanine) using Grunberg-Manago's enzyme system for making polynucleotides.

1962 - Max Perutz and John Kendrew share a Nobel prize for their work on the structure of hemoglobin and myoglobin.

1965 - Genetic code fully cracked through trial-and-error experimental work.

1966 - Kimishige Ishizaka discovers a new type of immunoglobulin, IgE, that develops allergy and explains the mechanisms of allergy at molecular and cellular levels.

1967 - John Gurden uses nuclear transplantation to clone an African clawed frog; first cloning of a vertebrate using a nucleus from a fully differentiated adult cell.

1968 - Fred Sanger uses radioactive phosphorus as a tracer to chromatographically decipher a 120 base long RNA sequence.

1969 - Dorothy Crowfoot Hodgkin discovers the three-dimensional structure of insulin.

1970 - Hamilton Smith and Daniel Nathans discover DNA restriction enzymes.

1970 - Howard Temin and David Baltimore independently discover reverse transcriptase enzymes.

1972 - Albert Eschenmoser and Robert Woodward synthesize vitamin B-12.

1972 - Stephen Jay Gould and Niles Eldredge propose an idea they call "punctuated equilibrium," which states that the fossil record is an accurate depiction of the pace of evolution, with long periods of "stasis" (little change) punctuated by brief periods of rapid change and species formation (within a lineage).

1972 - SJ Singer and GL Nicholson develop the fluid mosaic model, which deals with the make-up of the membrane of all cells.

1974 - Manfred Eigen and Manfred Sumper show that mixtures of nucleotide monomers and RNA replicase will give rise to RNA molecules which replicate, mutate, and evolve.

1974 - Leslie Orgel shows that RNA can replicate without RNA-replicase and that zinc aids this replication.

1977 - John Corliss, Jack Dymond, Louis Gordon, John Edmond, Richard von Herzen, Robert Ballard, Kenneth Green, David Williams, Arnold Bainbridge, Kathy Crane, and Tjeerd van Andel discover chemosynthetically based animal communities located around submarine hydrothermal vents on the Galapagos Rift.

1977 - Walter Gilbert and Allan Maxam present a rapid DNA sequencing technique which uses cloning, base destroying chemicals, and gel electrophoresis.

1977 - Frederick Sanger and Alan Coulson present a rapid gene sequencing technique which uses dideoxynucleotides and gel electrophoresis.

1978 - Frederick Sanger presents the 5,386 base sequence for the virus PhiX174; first sequencing of an entire genome.

1982 - Stanley B. Prusiner proposes the existence of infectious proteins, or prions. His idea is widely derided in the scientific community, but he wins a Nobel Prize in 1997.

1983 - Kary Mullis invents "PCR" (polymerase chain reaction), an automated method for rapidly copying sequences of DNA.

1984 - Alec Jeffreys devises a genetic fingerprinting method.

1985 - Harry Kroto, J.R. Heath, S.C. O'Brien, R.F. Curl, and Richard Smalley discover the unusual stability of the buckminsterfullerene molecule and deduce its structure.

1986 - Alexander Klibanov demonstrates that enzymes can function in non-aqueous environments.

1990-present

1990 - Napoli, Lemieux and Jorgensen discover RNA interference (1990) during experiments aimed at the color of petunias.

1990 - Wolfgang Krätschmer, Lowell Lamb, Konstantinos Fostiropoulos, and Donald Huffman discover that Buckminsterfullerene can be separated from soot because it is soluble in benzene.

1995 - Publication of the first complete genome of a free-living organism.

1996 - Dolly the sheep is first clone of an adult mammal.

2001 - Publication of the first drafts of the complete human genome.

2002 - First virus produced 'from scratch,' an artificial polio virus that paralyzes and kills mice.

4

Case Studies of Streams in Bioscience: Illustrations from Biophysics, Medicine & Neuroscience

BIOPHYSICS

Biophysics is an interdisciplinary science that applies the theories and methods of physics to questions of biology. Biophysics research today comprises a lot of specific biological studies, which don't share a unique identifying factor, nor subject themselves to clear and concise definitions. The studies included under the umbrella of biophysics range from sequence analysis to neural networks. In the past, biophysics included creating mechanical limbs and nanomachines to regulate biological functions. Today, these are more commonly referred to as belonging to the fields of bioengineering and nanotechnology respectively.

Biophysics is an interdisciplinary science that applies physical techniques for the study of biological phenomena. Traditional studies in biochemistry and molecular biology are conducted using statistical ensemble experiments, typically using pico- to micro-molar concentrations of macromolecules. Because the molecules that comprise living cells are so small, techniques such as PCR amplification, gel blotting, fluorescence labeling and in vivo staining are used so that experimental results are observable with an unaided eye or, at most, optical magnification. Using these techniques, researchers in these subjects attempt to elucidate the complex systems of interactions that give rise to the processes that make life possible.

Biophysics, in contrast, typically addresses similar biological questions, but the questions are asked at a molecular level. By drawing knowledge and experimental techniques from a wide variety of disciplines (as described below), biophysicists are able to indirectly observe or model the structures and interactions of *individual* molecules or complexes of molecules.

In addition to things like solving a protein structure or measuring the kinetics of interactions, biophysics is also understood to encompass research areas that apply models and experimental techniques derived from physics (*e.g.* electromagnetism and quantum mechanics) to larger systems such as tissues or organs (hence the inclusion of basic neuroscience as well as more applied techniques such as fMRI). Biophysics often does not have university-level departments of its own, but have presence as groups across departments within the fields of biology, biochemistry, chemistry, computer science, mathematics, medicine, pharmacology, physiology, physics, and neuroscience.

What follows is a list of examples of how each department applies its efforts toward the study of biophysics. This list is hardly all inclusive. Nor does each subject of study belong exclusively to any particular department. Each academic institution makes its own rules and there is much mixing between departments.

- Biology and molecular biology—Almost all forms of biophysics efforts are included in some biology department somewhere. To include some: gene regulation, single protein dynamics, bioenergetics, patch clamping, biomechanics.
- Structural biology—angstrom-resolution structures of proteins, nucleic acids, lipids, carbohydrates, and complexes thereof.
- Biochemistry and chemistry—biomolecular structure, siRNA, nucleic acid structure, structure-activity relationships.
- Computer science—Neural networks, Biomolecular and drug databases.
- Computational chemistry—Molecular dynamics simulation, Molecular docking, Quantum chemistry
- Bioinformatics—sequence alignment, structural alignment, Protein structure prediction
- Mathematics—graph/network theory, population modeling, dynamical systems, phylogenetics.
- Medicine and neuroscience—tackling neural networks experimentally (brain slicing) as well as theoretically (computer models), membrane permitivity, gene therapy, understanding tumors.
- Pharmacology and physiology—channel biology, biomolecular interactions, cellular membranes, polyketides.

- Physics—Biomolecular free energy, stochastic processes, covering dynamics.

Many biophysical techniques are unique to this field. Many of the research traditions in biophysics were initiated by scientists who were straight physicists, chemists, and biologists by training.

Main Topics of Biophysics and Related Fields

- Animal locomotion
- Cellular biophysics
- Channels, receptors and transporters
- Electrophysiology
- Cell membranes
- Bioenergetics
- Gravitational biology
- Molecular motors
- Muscle and contractility
- Nucleic acids
- Photobiophysics and biophotonics
- Proteins
- Signaling
- Supramolecular assemblies
- Spectroscopy, imaging, etc.
- Systems neuroscience
- Neural encoding
- Bionics
- Polysulphur membranes
- Biosensor and Bioelectronics
- Quantum biology

Techniques

Biophysical techniques are methods used for gaining information about biological systems on an atomic or molecular level. They overlap with methods from other branches of science.

- Electron microscopy, to gain high-resolution images of subcellular structures
- Dual Polarisation Interferometry, an analytical technique used to understand the real-time structure and behaviour of a wide range of molecular systems and interactions through quantitative measurement.
- Circular dichroism, a method for detecting chiral groups in molecules, especially to determine the secondary structure of proteins
- Fluorescence spectroscopy, which can be used to detect structural rearrangements, as well as interactions of biomolecules
- Force spectroscopy probes the mechanical properties of individual molecules or macromolecular assemblies using small flexible cantilevers, focussed laser light, or magnetic fields.
- Gel electrophoresis, which is used to determine the mass, the charge and the interactions of biological molecules
- Isothermal Titration Calorimetry or ITC which measure the heat effects caused by interactions
- Mass spectrometry is a technique that gives the molecular mass with great accuracy.
- Microscopy, for example using laser instruments for scanning and transmission.
- Optical tweezers and Magnetic tweezers allow for the manipulation of single molecules, providing information about DNA and its interaction with proteins and molecular motors, such as Helicase and RNA polymerase.
- NMR spectroscopy, giving information about the exact structure of biological molecules, as well as on dynamics
- Single molecule spectroscopy is a general term applied to a class of techniques that are sensitive enough to detect single molecules and often incorporates fluorescence detection.
- Small angle X-ray scattering (SAXS) is a technique that gives a rough low resolution molecular structure.
- Spectrophotometry, the measurement of the transmission of light through different solutions or substances at different wavelengths of light. Colorimetry is an example of this.
- Ultracentrifugation, which gives information on the shape and mass of molecules
- Various chromatography technique, which are used for the purification and analysis of biological molecules

- X-ray crystallography, another method to gain access to the exact structure of molecules with atomic resolution

Animal Locomotion

In biology and physics, animal locomotion is the study of how animals move, and is part of biophysics.

Much of the study is an application of Newton's third law of motion: if at rest, to move forwards an animal must push something backwards. Terrestrial animals must push the solid ground, swimming and flying animals must push against a fluid or gas (either water or air). The topic splits into the following disjoint categories:

1. animal locomotion in air (for example bird flight and insect flight)
2. animal locomotion on land (walking, hand walking and running)
3. animal locomotion in water (swimming including fish and ducks)
4. animal locomotion on the surface layer (small animals relying on surface tension such as the water strider)
5. animal locomotion by water-walkers (the Common Basilisk).
6. animal locomotion through the ground ('swimming' in sand or mud, or burrowing)
7. animal locomotion on the sea floor
8. animal locomotion on steep, vertical, and overhanging surfaces (climbing on trees or on rockfaces—for example, lemurs in trees, mountain goats on a cliff face, or flies or geckos on the ceiling)

The distinction between the second and third topics is that in the third, the animal does not need to expend energy to defeat gravity; in or on the water, buoyancy counteracts the animal's weight.

Terrestrial Locomotion in Animals

A number of animals have evolved so as to be able to travel over the ground. Terrestrial locomotion has evolved many times as animals moved onto the land from the water.

Types of Locomotion

There are three basic forms of locomotion found among terrestrial animals

- Legged—Moving by using appendages
- Slithering—moving using the bottom surface of the body
- Rolling—rotating the body over the substrate

Stance

Appendages can be used for movement in a number of ways. The stance, the way the body is supported by the legs, is an important aspect. Charig 1972 identified three main ways in which vertebrates support themselves with their legs—the sprawling stance, the semi-erect stance, and the fully erect stance. Some animals may use different stances in different circumstances, depending on the stance's mechanical and energetic advantages. The most basic is the sprawling stance. Here the legs are used to drag the body over the land. This is the earliest form of use of legs on land. Amphibious fish such as the mudskipper drag themselves across land on their sturdy fins. Many reptiles and amphibians, some or all of the time, use this method of locomotion. Among invertebrates there is anecdotal evidence that some octopus species (such as the *Pinnoctopus* genus), sometimes to pursue prey between rockpools, can also drag themselves across land a short distance by hauling its body along by it tentacles.

The second form of stance found among legged terrestrial animals is the semi-erect stance. Here the legs are to the side, but the body is held above the substrate. This mode of locomotion is found among some reptiles and amphibians. It is also the main stance of the crocodilians. A few mammals, such as the platypus also use this stance. Among the invertebrates most arthropods, which includes the most diverse group of animals—the insects, have a stance which might best be described as semi-erect.

Finally there is the main form of stance of mammal and birds, the fully erect stance. In these groups the legs are placed beneath the body. This is often linked with the evolution of endothermy (Bakker 1988). The fully erect stance is not necessarily the 'most-evolved' stance, evidence suggests that crocodilians evolved a semi-erect stance from ancestors with fully erect stance as a result of adapting to a mostly aquatic lifestyle (Reilly & Elias 1998). For example, the mesozoic prehistoric crocodilian *Erpetosuchus* is believed to have had a fully erect stance and been terrestrial.

Legged Locomotion

Movement on appendages is the most common form of terrestrial locomotion, it is the basic form of locomotion of two major groups with many terrestrial members, the vertebrates and the arthropods. Important aspects of legged locomotion are stance—the way the body is supported by the legs, the number of legs—from two to many, and the foot shape—broad, on toes, etc. There are also many gaits—ways of moving the legs in order to locomote—such as walking, running, or hopping

Number of Legs

The number of locomotory appendages varies much between animals, and sometimes the same animal may use different numbers of its legs in different circumstances. The best contender for unipedal movement is the springtail, which while typically hexapedal, hurls itself away from danger using its furcula, a tail-like forked rod that can be rapidly unfurled from the underside of its body.

A fair number of species move and stand on two legs, that is, are bipedal. The group that is exclusively bipedal is the birds, which have an alternating gait. There are also a number of bipedal mammals. Most bipedal mammals move by hopping—the macropods and various jumping rodents. Only a few mammals such as humans and the giant pangolin commonly show an alternating bipedal gait. Also cockroaches and some lizards may run on their two hind legs. Macropods such as kangaroos are the only example of tripedal movement. They have thick muscular tails and when moving slowly may alternate between resting their weight on their tails and their two hind legs.

With the exception of the birds, all terrestrial vertebrate groups are mostly quadrupedal—the mammals, the reptiles, and the amphibians usually move on four legs. There are many quadrupedal gaits. The most diverse group of animals on earth, the insects, are included in a larger taxon known as hexapods, most of which are hexapedal, walking and standing on six legs. Exceptions among the insects include praying mantises which are quadrapeds, their front two legs having been modified for grasping, and some kinds of insect larva who may have no legs (*e.g.* maggots) or additional prolegs (*e.g.* caterpillars).

Spiders and many of their relatives move on eight legs—are octopedal. However, some creatures move on many more legs. Terrestrial crustaceans may have a fair number—woodlice having fourteen legs. Also, as previously mentioned, some insect larvae such as caterpillars have up to six additional fleshy prolegs in addition to the six legs standard to insects. Some species of invertebrate have even more legs, the unusual velvet worm having stubby legs under the length of its body, with around several dozen pairs of legs. Centipedes have one pair of legs per body segment, with typically around 50 legs, but some species having over 200. The terrestrial animals with the most legs are the millipedes, relatives of the centipedes. They have two pairs of legs per body segment, with common species having between 80 and 400 legs overall. However, the rare species Illacme plenipes can have up to 750 legs. Animals with many legs typically move by waves of motion travelling over their legs.

Foot

The basic form of the vertebrate foot is similar. However where on the foot the animals weight is placed can vary greatly between vertebrates. Most vertebrates, the amphibians, the reptiles, and some mammals such as humans and bears are plantigrade, walking on the whole of the underside of the foot. Many mammals, such as cats and dogs are digitigrade, walking on their toes, the greater stride length allowing more speed. Digitigrade mammals are also often adept at quiet movement. Birds are also digitigrade. Some animals such as horses are unguligrade, walking on the tips of their toes, this even further increases their stride length and thus their speed. A few mammals are also known to walk on their knuckles, at least for their front legs. Knuckle-walking allows the foot (hand) to specialise for food gathering and/or climbing, as with the great apes and the extinct chalicotheres, or for swimming, as with the platypus. In animals where feet have evolved into functional hands, hand walking is also possible.

Gaits

Animals show a vast range of gaits, the order that they place and lift their appendages in locomotion. Walking is the most common gait, where some feet are on the ground at any given time, and found in almost all legged animals. Running is considered to occur when at some points in the gait all feet are off the ground. This can be found in many animals, and is a faster but more energetically costly form of locomotion. Animals will use different gaits for different speeds, terrain, and situations. For example horses show four natural gaits, the slowest horse gait is the walk, then there are three faster running gaits which, from slowest to fastest, are the trot, the canter, and the gallop. Animals may also have unusual gaits that are used occasionally, such as for moving sideways or backwards. For example, the main human gaits are bipedal walking and running, but they employ many other gaits occasionally, including a four-legged crawl in tight spaces. In walking, and for many animals running, the motion of legs on either side of the body alternates, *i.e.* is out of phase.

Other animals, such as a horse when galloping, or an inchworm, alternate between their front and back legs. An alternative to a gait which alternates between legs is hopping or saltation, where all legs move together. As a main means of locomotion, this is usually found in bipeds or semi-bipeds. Among the mammals saltation is commonly used among macropods (kangaroos and their relatives), jerboas, springhares, kangaroo rats, hopping mice, and gerbils. Certain tendons in kangaroo hind legs are very elastic, allowing kangaroos to effectively bounce along conserving energy from hop to hop, making hopping a very energy efficient way to move around in their nutrient poor environment. Saltation is also used by many small birds. Frogs and fleas also hop.

Most animals move in the direction of their head. However there are some exceptions. Crabs move sideways, and naked mole rats which live in tight tunnels underground can move backward or forward with equal facility.

Gait analysis is the study of gait in humans and other animals. This may involve videoing subjects with markers on particular anatomical landmarks and measuring the forces of their footfall using floor transducers (strain gauges). Skin electrodes may also be used to measure muscle activity.

Flying and Gliding Animals

A number of animals have evolved aerial locomotion, either by powered flight or by gliding. Flying and gliding animals have evolved separately many times, without any single ancestor. Flight has evolved at least four times, in the insects, pterosaurs, birds, and bats. Gliding has evolved on many more occasions. Usually the development is to aid canopy animals in getting from tree to tree, although there are other possibilities. Gliding, in particular, has evolved among rainforest animals, especially in the rainforests of Asia (most especially Borneo) where the trees are tall and quite widely spaced.

Types of Aerial Locomotion

- Falling: Decreasing altitude under the force of gravity, using no adaptions to increase drag or provide lift.
- Parachuting: Defined as falling at greater than 45 degrees from the horizontal with adaptations to increase drag forces. Very small animals may be carried up by the wind.
- Gliding: Defined as falling at less than 45 degrees from the horizontal. Lift caused by some kind of aerofoil mechanism, allowing slowly falling directed horizontal movement. Streamlined to decrease drag forces to aid aerofoil. Often some maneuverability in air. Gliding animals have a lower aspect ratio (wing length/wing breadth) than flyers.
- Flying: Flapping of wings to produce thrust. May ascend without the aid of the wind, as opposed to gliders and parachuters.
- Soaring: Appears similar to gliding but is actually very different, requiring specific physiological and morphological adaptations. The animal keeps aloft on rising warm air (thermals) without flapping its wings. Only large animals can be efficient soarers.

These forms of aerial locomotion are not mutually exclusive and indeed many animals will employ two or more of the methods. Two other common forms of aerial locomotion for humans that are not employed in the rest of the animal kingdom are heli-propulsion and the balloon.

Ecology of Aerial Locomotion

Although only four groups of animal have evolved flight, all three extant groups are very successful, suggesting that flight is a very successful strategy once evolved.

Bats, after rodents, have the most species of any mammalian order, about 20% of all mammalian species. Birds have the most species of any class of terrestrial vertebrates. Finally insects have more species than all other animal groups combined.

Flying animals may have evolved from gliding animals. However gliding is not necessarily just an evolutionary route to flying and has some advantages of its own. Gliding is a very energy efficient way of travelling from tree to tree. An argument made is that many gliding animals eat low energy foods such as leaves and are restricted to gliding because of this, whereas flying animals eat more high energy foods such as fruits, nectar, and insects. In contrast to flight, gliding has evolved independently many times (more than a dozen times among extant vertebrates), however these groups have not radiated nearly as much as have groups of flying animals.

One point of interest is the distribution of gliding animals. Many gliding animals are found in Southeast Asia, some in Africa, and there are no gliding vertebrates in South America. However, many more animals in South America have prehensile tails than in Africa and Southeast Asia. It has been argued that gliding animals dominate in Southeast Asia as the forests are less dense than in South America. In dense forest there is not room to glide, but a prehensile tail is very useful for moving from tree to tree. Also South American rainforests tend to have more lianas as there are fewer large animals to eat them compared to Africa and Asia; these lianas would aid climbers but obstruct gliders. Curiously Australia contains many mammals with prehensile tails and also many mammals which can glide, in fact all Australian mammalian gliders have tails that are prehensile to an extent.

Only a few animals are known to have specialised in soaring, the larger of the extinct pterosaurs, and some large birds. Powered flight is very energetically expensive for large animals, but for soaring their size is an advantage as it allows them a low wing loading, that is a large wing areas relative to their weight, which maximizes lift. Soaring is very energetically efficient.

Biomechanics of Aerial Locomotion

The forms of aerial locomotion for which the biomechanics are most studied are bird flight and insect flight. The UCMP exhibit on vertebrate flight contains a broad introduction to the biomechanics of flying and gliding vertebrates.

Extremes and Limits

Flying / Soaring

- Fastest. The fastest of all known flying animals is the peregrine falcon, which when diving has been recorded flying at 300 km/h or faster. The fastest animal in flapping flight might be the white-throated needle-tailed swift, at 170 km/h. In level flapping flight, a good contender for the fastest living animal recorded is the red-breasted merganser at 100 mi/h (160 km/h).

- Smallest. There is no real minimum size for getting airborne. Indeed, there are many bacteria floating in the atmosphere that constitute part of the aeroplankton. However, to move about under one's own power and not be overly affected by the wind requires a certain amount of size.
- Largest. The largest known flying animal was formerly thought to be *Pteranodon*, with a wingspan of up to 7.5 m. However, the more recently discovered *Quetzalcoatlus* is much larger, with estimates of the wingspan ranging from 9 m to 18 m. Although it is widely thought that Quetzalcoatlus reached the size limit of a flying animal, it should be noted that the same was once said of Pteranodon. The heaviest living flying animal is the great bustard at 21 kg. The wandering albatross has the greatest wingspan of any living flying animal at 3.63 m (11 ft 11 in). Among living animals which fly over land, the Andean condor and the marabou stork have the largest wingspan at 3.2 m.
- Slowest. Most flying animals need to travel forward at a minimum speed to stay aloft. However, some creatures can stay in the same spot, known as hovering, either by rapidly flapping the wings, as do hummingbirds, hoverflies, dragonflies, and some others, or carefully using thermals, as do some birds of prey. The slowest flying non-hovering bird recorded is the American woodcock, at 8 km/h. However, many insects probably fly much slower than this.
- Highest flying. There are records of a Rueppell's Vulture *Gyps rueppelli*, a large vulture, being sucked into a jet engine 37,900 feet (11,550 m) above the Ivory Coast in West Africa. The animal that flies highest most regularly is the bar-headed goose *Anser indicus*, which migrates directly over the Himalayas between its nesting grounds in Tibet and its winter quarters in India. They are sometimes seen flying well above the peak of Mount Everest at 8,848 m (29,028 feet).
- Most maneuverable. A number of flying animals are known for their maneuverability. Many animals that can hover are often very maneuverable, being able to move in any direction as well as stay still. Other flying animals known for their aerial acrobatics are bats and crows.

Gliding / Parachuting

- Most efficient glider. This can be taken as the animal that moves most horizontal distance per metre fallen. Possible candidates are the flying squirrels which are known to glide up to 200 m and flying fish has been observed to glide for hundreds of meters on the drafts on the edge of waves with only their initial leap from the water to provide height.

- Most maneuverable glider. Paradise tree snakes, Chinese gliding frogs, and gliding ants have all been observed as having considerable capacity to turn in the air. Many other gliding animals may also be able to turn, but which is the most maneuverable is difficult to assess.
- Most efficient parachuter. This could be the animal that is the slowest falling, or the animal that is slowest falling given its weight.

Animals which parachute, glide, or fly (living)

Vertebrates

Amphibians

- Rhacophoridae flying frogs (gliding). Gliding has evolved independently in two families of tree frogs, the Old World Rhacophoridae and the New World Hylidae. Within each lineage there are a range of gliding abilities from non-gliding, to parachuting, to full gliding. A number of the Rhacophoridae have adaptation for gliding, the main feature being enlarged toe membranes. For example, the Malayan flying frog glides using the membranes between the toes of its limbs, and small membranes located at the heel, the base of the leg, and the forearm. Some of the frogs are quite accomplished gliders, for example, the Chinese gliding frog *Polypedates dennysi* can maneuver in the air, making two kinds of turn, either rolling into the turn (a banked turn) or yawing into the turn (a crabbed turn).
- Hylidae flying frogs (gliding). The other frog family that contains gliders.

Fish

- Flying fish (gliding). There are over 50 species of flying fish belonging to the family Exocoetidae. They are mostly marine fishes of small to medium size. The largest flying fish can reach lengths of 45 cm, but most species measure less than 30 cm in length. They can be divided into two-winged varieties and four-winged varieties. The glides are usually up to 30-50 meters in length, but some have been observed soaring for hundreds of metres using the updraft on the leading edges of waves. The fish can also make a series of glides, each time dipping the tail into the water to produce forward thrust. It has been suggested that the genus *Exocoetus* is on an evolutionary borderline between flight and gliding. It flaps its enlarged pectoral fins when airborne, but still seems only to glide, as there is no hint of a power stroke.

- Hemirhamphid half-beaks (gliding). A group related to the Exocoetidae, one or two hemirhamphid species possess enlarged pectoral fins and show true gliding flight rather than simple leaps. Marshall (1965) reports that Euleptorhamphus viridis can cover 50 m in two separate hops.
- Freshwater butterflyfish (possibly gliding). *Pantodon buchholzi* has the ability to jump and possibly glide a short distance. It can move through the air several times the length of its body. While it does this, the fish flaps its large pectoral fins, giving it its common name. However, it is debated whether the freshwater butterfly fish can truly glide, Saidel et al (2004) argue that it cannot.
- Freshwater hatchetfish (possibly flying). There are 9 species of freshwater hatchetfish split among 3 genera. Freshwater hatchetfish have an extremely large sternal region that is fitted with a large amount of muscle that allows it to flap their pectoral fins. They can move in a straight line over a few meters to escape predators.

Birds

- Birds (flying, soaring) Again the species are too numerous to nominate. Bird flight is one of the most studied forms of aerial locomotion in animals. See List of soaring birds for birds that can soar as well as fly.

Reptiles

- *Draco* lizards (gliding). There are 28 species of lizard of the genus *Draco*, found in Sri Lanka, India, and Southeast Asia. They live in trees, feeding on tree ants, but nest on the forest floor. They can glide for up to 100 m, but usually only glide up to 20-30 m between trees as forest trees are often not so widely spaced. Unusually, their patagium (gliding membrane) is supported on elongated ribs rather than the more common situation among gliding vertebrates of having the patagium attached to the limbs. When extended, the ribs form a semi-circle on either side the lizards body and can be folded to the body like a folding fan.
- Gliding Lacertids (gliding). There are two species of gliding lacertid, of the genus *Holaspis*. Found in Africa. They have fringed toes and tail sides and can flatten their bodies for gliding.
- *Ptychozoon* flying geckos (gliding). There are six species of gliding gecko, of the genus *Ptychozoon*, from Southeast Asia. These lizards have small flaps of skin along their limbs, torso, tail, and head that catch the air and enable them to glide.

- *Cosymbotus* flying gecko (gliding). Similar adaptations to Ptychozoon are found in the two species of the gecko genus *Cosymbotus*.
- *Chrysopelea* snakes (gliding/parachuting). Five species of snake from Southeast Asia, Melanesia, and India. The paradise tree snake of southern Thailand, Malaysia, Borneo, Philippines, and Sulawesi is the most capable glider of those snakes studied. It glides by stretching out its body sideways by opening its ribs so the belly is concave, and by making lateral slithering movements. It can remarkably glide up to 100 m and make 90 degree turns. Follow this link for videos of gliding snakes.

Reptiles

- Extinct reptiles similar to *Draco* (gliding). There are a number of unrelated extinct lizard-like reptiles with similar "wings" to the *Draco* lizards. *Icarosaurus*, *Daedalosaurus*, *Coelurosauravus*, *Weigeltosaurus*, and *Kuehneosaurus*. The largest of these, *Kuehneosaurus*, has a wingspan of 30 cm, and was estimated to be able to glide about 30 m.
- *Sharovipteryx* (gliding). This strange reptile, sometimes proposed as a pterosaur ancestor, from the Upper Triassic of Kirghiia unusually had a membrane on its elongated hind limbs, as opposed to the forelimbs, which is much more usual. In some reconstructions they had webbing on the forelimbs and neck as well.
- *Longisquama insignis* (possibly gliding/parachuting). This small reptile may have had long paired feather-like scales on its back, however it has been more recently argued that the scales form just a single dorsal frill. If paired, they may have been used for parachuting.
- Pterosaurs (flying). Pterosaurs were the first flying vertebrates, and are generally agreed to have been sophisticated flyers. They had large wings formed by a patagium stretching from the torso to a dramatically lengthened fourth finger. There were hundreds of species, most of which are thought to have been intermittent flappers, and many soarers. The largest known flying animals are pterosaurs.

Mammals

- Sugar Glider wrist winged gliders found in Australia, New Guinea, and Indonesia. In order to glide the animal jumps form a high platform and extends its arms and legs to expose its gliding membrane.
- Flying phalangers or wrist-winged gliders (subfamily Petaurinae) (gliding). Marsupials found in Australia, New Guinea, and Borneo. The gliding membranes are hardly noticeable until they jump. On jumping, the animal

extends all four legs and stretches the loose but muscularly controlled folds of skin. The subfamily contains seven species. Of the six species in the genus Petaurus, the Sugar glider and the Biak glider are the most common species. The lone species in the genus Gymnobelideus, Leadbeater's Possum has only a vestigial gliding membrane.

- Greater glider (*Petauroides volans*) (gliding). The only species of the genus Petauroidae of the family Pseudocheiridae. This Marsupial is found in Australia, and was originally classed with the flying phalangers, but is now recognised as separate. Flying membrane only extends as far as elbow, rather than to wrist as in Petaurinae.

- Feather-tailed possums (family Acrobatidae) (gliding). This family of Marsupials contains two genera, each with one species. The Feathertail Glider (*Acrobates pygmaeus*), found in Australia is the size of a very small mouse and is the smallest mammalian glider. The Feather-tail Possum (*Distoechurus pennatus*) is found in New Guinea. Both species have a stiff-haired feather-like tail that helps it steer in the air.

- Bats (flying). There are many species of bat, again too numerous to nominate.
- Anomalure or scaly-tailed flying squirrels (Anomaluridae family) (gliding). These brightly coloured African rodents are not squirrels but have evolved to a resemble flying squirrels by convergent evolution. There are seven species, divided in three genera. All but one species has gliding membranes between their front and hind legs. One genus is particularly small and is known as flying mice, but similarly they are not mice.
- Colugos or Flying lemurs (order Dermoptera) (gliding). There are two species of flying lemur. This is not a lemur, which is a primate, but molecular evidence suggests that colugos are a sister group to primates, however some mammologists suggest they are a sister group to bats. Found in Southeast Asia, the colugo is probably the mammal most adapted for gliding, with a patagium that is as large as geometrically possible. They can glide as far a 70 m with minimal loss of height.

- Sifaka and possibly some other primates (possible limited gliding/parachuting). A number of primates have been suggested to have adaptations that allow limited gliding and/or parachuting; sifikas, indris, galagos and saki monkeys. Most notably the sifaka, a type of lemur, has thick hairs on its forearms that have been argued to provide drag, and a small membrane under its arms that has been suggested provide lift by having aerofoil properties.

- Cats and maybe others. (very limited parachuting). If they fall cats spread their bodies to maximise drag, a very limited form of parachuting. Cats have an innate 'righting reflex' that allows them to rotate their bodies so they fall feet first. Some other animals may show similar very limited parachuting. There are also anecdotal accounts of less limited parachuting, or even semi-gliding, in palm civets.

- Flying squirrels (subfamily Petauristinae) (gliding). There are 43 species divided between 14 genera of flying squirrel. Flying squirrels are found almost worldwide in tropical (Southeast Asia, India, and Sri Lanka), temperate, and even Arctic environments. They tend to be nocturnal. When a flying squirrel wishes to cross to a tree that is further away than the distance possible by jumping, it extends the cartilage spur on its elbow or wrist. This opens out the flap of furry skin (the patagium) that stretches from its wrist to its ankle. It glides spread-eagle and with its tail fluffed out like a parachute, and grips the tree with its claws when it lands. Flying squirrels have been reported to glide over 200 m.

Birds

- Theropods (gliding/flying). There were several species of theropod dinosaur thought to capable of gliding or flying, that are not classified as birds (though they are closely related). Some species (*Microraptor gui*, *Microraptor zhaoianus*, and *Cryptovolans pauli*) have been found that were fully feathered on all four limbs, giving them four 'wings' that they are believed to have used for gliding or flying.

Mammals

- *Volaticotherium antiquum* (gliding). The earliest known flying or gliding mammal. This squirrel-sized animal belonged to a now extinct ancestral line and was not related to modern day flying or gliding mammals, such as bats or gliding marsupials. It lived at least 125 million years ago and used a fur-covered skin membrane to glide through the air.

Invertebrates

- Gliding ants (gliding). These flightless insects have secondarily gained some capacity to move through the air. Gliding has evolved independently in a number of arboreal ant species from the groups Cephalotini, Pseudomyrmecinae, and Formicinae (mostly *Camponotus*). All arboreal dolichoderines and non-cephalotine myrmicines except *Daceton armigerum* do not glide. Living in the rainforest canopy like many other gliders, gliding ants use their gliding to return to the trunk of the tree they live on should they fall or be knocked off a branch. Gliding was first discovered for *Cephalotes atreus* in the Peruvian rainforest. *Cephalotes atreus* can make 180 degree turns, and locate the trunk using visual cues, succeeding in landing 80% of the time. Unique among gliding animals, Cephalotini and Pseudomyrmecinae ants glide abdomen first, the Forminicae however glide in the more conventional head first manner. The following page has some good videos of gliding ants.

- Insects (flight). The first of all animals to evolve flight, insects are also the only invertebrates that have evolved flight. The species are too numerous to list here. Insect flight has been studied in some detail, but less than bird flight.
- Spiders (parachuting). The young of some species of spiders travel through the air by using silk draglines to catch the wind, as may some smaller species of adult spider, such the money spider family. This behavior is commonly known as "ballooning". Ballooning spiders make up part of the aeroplankton.
- Flying squid (gliding). Several oceanic squids, such as the Pacific flying squid, will leap out of the water to escape predators, an adaptation similar to that of flying fish. Smaller squids will fly in shoals, and have been observed to cover distances as long as 50 meters. Small fins towards the back of the mantle do not produce much lift, but do help stabilize the motion of flight. They exit the water by expelling water out of their funnel, indeed some squid have been observed to continue jetting water while airborne possibly providing thrust even after leaving the water. This may make flying squid the only animals with, at a push, jet-propelled aerial locomotion.

Animal Locomotion on the Surface Layer of Water

Animal locomotion on the surface layer is the study of animal locomotion in the case of small animals that live on the surface layer of water, relying on surface tension to stay afloat. There are two means of walking on water; the regime determined by the ratio of the animal's weight to the maximum vertical force that the surface layer can extert. Creatures such as the basilisk lizard have a weight which is larger than the surface tension can support and are discussed in animal locomotion by water-walkers. Surface living animals such as the water strider typically have hydrophobic feet covered in small hairs that prevent the feet from breaking the surface and becoming wet.

Water striders in particular are known to have feet that have minute hairs on them. According to biophysicist David L. Hu, there are at least 342 species of water striders. As striders increase in size, their legs become proportionately longer, with Gigantometra gigas having a length of over 20 cm and a weight of about 40 millinewtons. Water striders generate thrust by shedding vortices in the water: a series of "U"-shaped vortex filiaments is created during the power stroke. The two free ends of the "U" are attached to the water surface. These vortices transfer enough (backward) momentum to the water to propel the animal forwards (note that some momentum is transferred by capillary waves; see Denny's paradox for a more detailed discussion)

Meniscus Climbing

To pass from the water surface to land, a water-walking insect must contend with the slope of the meniscus at the water's edge. Many such insects are unable to climb this meniscus using their usual propulsion mechanism. David Hu and coworker John W. M. Bush have shown that such insects climb meniscuses by assuming a fixed body posture. This deforms the water surface and generates capillary forces that propels the insect up the slope without moving its appendages. Hu and Bush conclude that meniscus climbing is an unusual means of propulsion in that the insect propels itself in a quasi-static configuration; without moving its appendages. Biolocomotion is generally characterized by the transfer of muscular strain energy to the kinetic and gravitational potential energy of the creature, and the kinetic energy of the suspending fluid. In contrast, meniscus climbing has a different energy pathway: by deforming the free surface, the insect converts muscular strain to the surface energy that powers its ascent.

Climbing Animals

There are a diverse range of climbing animals; animals that spend much of their time moving on steep, vertical, or overhanging surfaces and have appropriate adaptations for such locomotion. Climbing animals can be roughly divided into two groups. Those animals which move steep, vertical, or overhanging rock surfaces—rockface locomotion. The other group are those who move among tall vegetation—arboreal locomotion. These two environments may produce quite different methods of climbing. However, in some cases the climbing methods are similar, especially for small animals for which a rockface and a tree trunk may present similar problems for locomotion.

Ion Channel

Ion channels are pore-forming proteins that help to establish and control the small voltage gradient across the plasma membrane of all living cells by allowing the flow of ions down their electrochemical gradient. They are present in the membranes that surround all biological cells.

Basic Features

An ion channel is an integral membrane protein or more typically an assembly of several proteins. Such "multi-subunit" assemblies usually involve a circular arrangement of identical or homologous proteins closely packed around a water-filled pore through the plane of the membrane or lipid bilayer. The pore-forming subunit(s) are called the á subunit, while the auxiliary subunits are denoted â, ã, and so on. While some channels permit the passage of ions based solely on charge, the archetypal channel pore is just one or two atoms wide at its narrowest point. It conducts a specific species of ion, such as sodium or potassium, and conveys them through the membrane single file—nearly as quickly as the ions move through free fluid. In some ion channels, passage through the pore is governed by a "gate," which may be opened or closed by chemical or electrical signals, temperature, or mechanical force, depending on the variety of channel.

Biological Role

Because "voltage-gated" channels underlie the nerve impulse and because "transmitter-gated" channels mediate conduction across the synapses, channels are especially prominent components of the nervous system. Indeed, most of the offensive and defensive toxins that organisms have evolved for shutting down the nervous systems of predators and prey (*e.g.*, the venoms produced by spiders, scorpions, snakes, fish, bees, sea snails and others) work by plugging ion channel pores. In addition, ion channels figure in a wide variety of biological processes that involve rapid changes in cells, such as cardiac, skeletal, and smooth muscle contraction, epithelial transport of nutrients and ions, T-cell activation and pancreatic beta-cell insulin release. In the search for new drugs, ion channels are a favorite target.

Diversity

- Voltage-gated sodium channels: Like other voltage-gated channels, these channels open and close in response to membrane potential. This family contains at least 9 members and is largely responsible for action potential creation and propagation. The pore-forming á subunits are very large (up to 4,000 amino acids) and consist of four homologous repeat domains (I-IV) each comprising six transmembrane segments (S1-S6) for a total of 24 transmembrane segments. The members of this family also coassemble with auxiliary â subunits, each spanning the membrane once. Both á and â subunits are extensively glycosylated.

- Voltage-gated calcium channels: As with the other voltage-gated channels, these open and close according to the membrane potential. This family contains 10 members, though these members are known to coassemble with $á_2$ä, â, and ã subunits. These channels play an important role in both linking muscle excitation with contraction as well as neuronal excitation with transmitter release. The á subunits have an overall structural resemblance to those of the sodium channels and are equally large.

- Potassium channels: This superfamily is comprised of four families of channels, which are grouped based on homology and activation. Potassium channels are near ubiquitous in their expression and are primarily permeable to potassium over other ions.

- Voltage-gated potassium channels: Like other voltage-gated channels, these K_V channels open and close according to membrane potential. This family contains almost 40 members, which are further divided into 12 subfamilies. These channels are known mainly for their role in repolarizing the cell membrane following action potentials. The á subunits have six transmembrane segments, homologous to a single domain of the sodium channels. Correspondingly, they assemble as tetramers to produce a functioning channel.
- Calcium-activated potassium channels: This family of channels is, for the most part, activated by intracellular Ca^{2+} and contains 8 members.
- Inward-rectifier potassium channels: These channels allow potassium to flow into the cell in an inwardly rectifying manner, i.e, potassium flows effectively into, but not out of, the cell. This family is composed of 15 official and 1 unofficial members and is further subdivided into 7 subfamilies based on homology. These channels are affected by intracellular ATP, PIP_2, and G-protein ââ subunits. They are involved in important physiological processes such as the pacemaker activity in the heart, insulin release, and potassium uptake in glial cells. They contain only two transmembrane segments, corresponding to the core pore-forming segments of the K_V and K_{Ca} channels. Their á subunits form tetramers.
- Two-pore-domain potassium channels: This family of 15 members form what is known as leak channels, and they follow Goldman-Hodgkin-Katz (open) rectification.
- Chloride channels: This superfamily of poorly understood channels consists of approximately 13 members.
- Transient receptor potential channels: This group of channels, normally referred to simply as TRP channels, is named after their role in Drosophila phototransduction. This family, containing at least 28 members, is incredibly diverse in its method of activation. Some TRP channels seem to be constitutively open, while others are gated by voltage, intracellular Ca^{2+}, pH, redox state, osmolarity, and mechanical stretch. These channels also vary according to the ion(s) they pass, some being selective for Ca^{2+} while others are less selective, acting as cation channels. This family is subdivided into 6 subfamilies based on homology: classical (TRPC), vanilloid receptors (TRPV), melastatin (TRPM), polycystins (TRPP), mucolipins (TRPML), and ankyrin transmembrane protein 1 (TRPA).

- Cyclic nucleotide-gated channels: This superfamily of channels contains two families: the cyclic nucleotide-gated (CNG) channels and the hyperpolarization-activated, cyclic nucleotide-gated (HCN) channels. It should be noted that this grouping is functional rather than evolutionary.
- Cyclic nucleotide-gated channels: This family of channels is characterized by activation due to the binding of intracellular cAMP or cGMP, with specificity varying by member. These channels are primarily permeable to monovalent cations such as K^+ and Na^+. They are also permeable to Ca^{2+}, though it acts to close them. There are 6 members of this family, which is divided into 2 subfamilies.
- Hyperpolarization-activated, cyclic nucleotide-gated channels: While these channels are voltage-gated, their opening is due to hyperpolarization rather than the depolarization required for other like channels. These channels are also sensitive to the cyclic nucleotides cAMP and cGMP, which alter the voltage sensitivity of the channel's opening. These channels are permeable to the monovalent cations K^+ and Na^+. There are 4 members of this family, all of which form tetramers of six-transmembrane á subunits. As these channels open under hyperpolarizing conditions, they function as pacemaking channels in the heart, particularly the SA node.
- Cation channels of sperm: This small family of channels, normally referred to as Catsper channels, is related to the two-pore channels and distantly related to TRP channels.
- Two-pore channels: This small family of 2 members putatively forms cation-selective ion channels. They are predicted to contain two K_V-style six-transmembrane domains, suggesting they form a dimer in the membrane. These channels are related to catsper channels channels and, more distantly, TRP channels.
- Light-gated channels like channelrhodopsin are directly opened by the action of light.
- *Ligand-gated* channels (LGICs): Also known as ionotropic receptors, this group of channels open in response to specific ligand molecules binding to the extracellular domain of the receptor protein. Ligand binding causes a conformational change in the structure of the channel protein that ultimately leads to the opening of the channel gate and subsequent ion flux across the plasma membrane. Examples of LGICs include the cation-permeable "nicotinic" Acetylcholine receptor, ionotropic glutamate-gated receptors and ATP-gated P2X receptors, and the anion-permeable ã-aminobutyric acid-gated $GABA_A$ receptor.

Detailed Structure

Channels differ with respect to the ion they let pass (for example, Na^+, K^+, Cl^-), the ways in which they may be regulated, the number of subunits of which they are composed and other aspects of structure. Channels belonging to the largest class, which includes the voltage-gated channels that underlie the nerve impulse, consists of four subunits with six transmembrane helices each. On activation, these helices move about and open the pore. Two of these six helices are separated by a loop that lines the pore and is the primary determinant of ion selectivity and conductance in this channel class and some others. The existence and mechanism for ion selectivity was first postulated in the 1960s by Clay Armstrong. The channel subunits of one such other class, for example, consist of just this "P" loop and two transmembrane helices. The determination of their molecular structure by Roderick MacKinnon using X-ray crystallography won a share of the 2003 Nobel Prize in Chemistry.

Because of their small size and the difficulty of crystallizing integral membrane proteins for X-ray analysis, it is only very recently that scientists have been able to directly examine what channels "look like." Particularly in cases where the crystallography required removing channels from their membranes with detergent, many researchers regard images that have been obtained as tentative. An example is the long-awaited crystal structure of a voltage-gated potassium channel, which was reported in May 2003. The detailed 3D structure of the magnesium channel from bacteria can be seen here. One inevitable ambiguity about these structures relates to the strong evidence that channels change conformation as they operate (they open and close, for example), such that the structure in the crystal could represent any one of these operational states. Most of what researchers have deduced about channel operation so far they have established through electrophysiology, biochemistry, gene sequence comparison and mutagenesis.

Diseases of Ion Channels

There are a number of chemicals and genetic disorders which disrupt normal functioning of ion channels and have disastrous consequences for the organism. Genetic disorders of ion channels and their modifiers are known as Channelopathies. See Category:Channelopathy for a full list.

Chemicals

- Tetrodotoxin (TTX), used by puffer fish and some types of newts for defense. It is a sodium channel blocker.
- Saxitoxin, produced by a dinoflagellate also known as red tide. It blocks voltage dependent sodium channels.
- Conotoxin, which is used by cone snails to hunt prey.

- Lidocaine and Novocaine belong to a class of local anesthetics which block sodium ion channels.
- Dendrotoxin is produced by mamba snakes which blocks potassium channels.

Genetic

- Shaker gene mutations cause a defect in the voltage gated ion channels, slowing down the repolarization of the cell.
- Equine hyperkalaemic periodic paralysis as well as Human hyperkalaemic periodic paralysis (HyperPP) are caused by a defect in voltage dependent sodium channels.
- Paramyotonia congenita (PC) and potassium aggravated myotonias (PAM)
- Generalized epilepsy with febrile seizures plus (GEFS+)
- Episodic Ataxia (EA)
- Familial hemiplegic migraine (FHM)
- spinocerebellar ataxia type 13
- Long QT syndrome is a ventricular arrhythmia syndrome caused by mutations in one or more of presently ten different genes, most of which are potassium channels and all of which affect cardiac repolarization.
- Brugada syndrome is another ventricular arrhythmia caused by voltage-gated sodium channel gene mutations.
- Cystic fibrosis is caused by mutations in the CFTR gene, which is a chloride channel.

The existence of ion channels was hypothesized by the British biophysicists Alan Hodgkin and Andrew Huxley as part of their Nobel Prize-winning theory of the nerve impulse, published in 1952. The existence of ion channels was confirmed in the 1970s with an electrical recording technique known as the "patch clamp," which led to a Nobel Prize to Erwin Neher and Bert Sakmann, the technique's inventors. Hundreds if not thousands of researchers continue to pursue a more detailed understanding of how these proteins work. In recent years the development of automated patch clamp devices helped to increase the throughput in ion channel screening significantly.

Rockface Locomotion

Balancing

A number of animals move on rockface which are steep or even near vertical by careful balancing and leaping. Perhaps the most exceptional of these are the various types of mountain dwelling caprid such as the Barbary sheep, markhor, tur, ibex, tahr, rocky mountain goat, and chamois. Their adaptation may include a soft rubbery pad between their hooves for grip, hooves with sharp keratin rims for lodging in small footholds, and prominent dew claws. The snow leopard, being a predator of such mountain caprids is also a spectacular balancer and leaper, being able to leap up to ~17m (~50ft). Other balancer and leapers include the mountain zebra, mountain tapir, and hyraxes.

Brachiating

Brachiation is arguably the epitomy of arboreal locomotion, and involved swinging with the arms from one handhold to another. Only a few species are brachiators, and all of these are primates; it is a major means of locomotion among spider monkeys and gibbons, and is occaisonaly used by female orangutans. Gibbons are the experts of this mode of locomotion, swinging from branch to branch distances of up to 15m (50ft), and traveling at speeds of as much as 56 km/h (35 mph).

Transmembrane Receptor

Transmembrane receptors are integral membrane proteins, which reside and operate typically within a cell's plasma membrane, but also in the membranes of some subcellular compartments and organelles. Binding to a signalling molecule or sometimes to a pair of such molecules on one side of the membrane, transmembrane receptors initiate a response on the other side. In this way they play a unique and important role in cellular communications and signal transduction.

Many transmembrane receptors are composed of two or more protein subunits which operate collectively and may dissociate when ligands bind, fall off, or at another stage of their "activation" cycles. They are often classified based on their molecular structure, or because the structure is unknown in any detail for all but a few receptors, based on their hypothesized (and sometimes experimentally verified) membrane topology. The polypeptide chains of the simplest are predicted to cross the lipid bilayer only once, while others cross as many as seven times (the so-called G-protein coupled receptors).

Domains

Like any integral membrane protein, a transmembrane receptor may be subdivided into three parts or *domains*.

The Extracellular Domain

The extracellular domain is the part of the receptor that sticks out of the membrane on the outside of the cell or organelle. If the polypeptide chain of the receptor crosses the bilayer several times, the external domain can comprise several "loops" sticking out of the membrane. By definition, a receptor's main function is to recognize and respond to a specific ligand, for example, a neurotransmitter or hormone (although certain receptors respond also to changes in transmembrane potential), and in many receptors these ligands bind to the extracellular domain.

The Transmembrane Domain

In the majority of receptors for which structural evidence exists, transmembrane alpha helices make up most of the transmembrane domain. In certain receptors, such as the nicotinic acetylcholine receptor, the transmembrane domain forms a protein-lined pore through the membrane, or ion channel. Upon activation of an extracellular domain by binding of the appropriate ligand, the pore becomes accessible to ions, which then pass through. In other receptors, the transmembrane domains are presumed to undergo a conformational change upon binding, which exerts an effect intracellularly. In some receptors, such as members of the 7TM superfamily, the transmembrane domain may contain the ligand binding pocket (evidence for this and for much of what else is known about this class of receptors is based in part on studies of bacteriorhodopsin, the detailed structure of which has been determined by crystallography).

The Intracellular Domain

The intracellular (or cytoplasmic) domain of the receptor interacts with the interior of the cell or organelle, relaying the signal. There are two fundamentally different ways for this interaction:

- With enzyme-linked receptors, the intracellular domain has *enzymatic activity*. Often, this is a tyrosine kinase activity. The enzymatic activity can also be located on an enzyme associated with the intracellular domain.
- The intracellular domain communicates via specific protein-protein-interactions with *effector proteins*, which in turn send the signal along a signal chain to its destination.

Regulation of Receptor Activity

There are several ways for the cell to regulate the activity of a transmembrane receptor. Most of them work through the intracellular domain. The most important ways are phosphorylation and internalization.

Electrophysiology

Electrophysiology is the study of the electrical properties of biological cells and tissues. It involves measurements of voltage change or electrical current flow on a wide variety of scales from single ion channel proteins, to whole tissues like the heart. In neuroscience, it includes measurements of the electrical activity of neurons, and particularly action potential activity.

Definition and Scope

Classical electrophysiologic techniques: Classical electrophysiology involves placing electrodes into various preparations of biologic tissue. The principle types of electrodes are: 1) simple solid conductors, such as discs and needles (singles or arrays), 2) tracings on a printed circuit boards, and 3) hollow tubes filled with an electrolyte, such as glass pippettes. The principal preparations include 1) living organisms, 2) excised tissue (acute or cultured), 3) dissociated cells from excised tissue (acute or cultured), 4) artificially grown cells or tissues, or 5) hybrids of the above.

If an electrode is small enough in diameter (on the order of microns), then the electrophysiologist may choose to insert the tip into a single cell. Such a configuration allows direct observation and recording of the intracellular electrical activity of a single cell. However, at the same time such invasive setup reduces the life of the cell. Intracellular activity may also be observed using a specially formed (hollow) glass pipette. In this technique, the microscopic pipette tip is pressed against the cell membrane, to which it tightly adheres.

The electrolyte within the pipette may be brought into fluid continuity with the cytoplasm by delivering pulse of pressure to the electrolyte in order to rupture the small patch of membrane encircled by the pipette rim, (whole cell recording). Alternatively, ionic continuity may be established by "perforating" the patch by allowing exogenous ion channels within the electrolyte to insert themselves into the membrane patch perforated patch recording. Finally, the patch may be left intact patch recording.

The electrophysiologist may choose not to insert the tip into a single cell. Instead, the electrode tip may be left in continuity with the extracellular space. If the tip is small enough, such a configuration may allow indirect observation and recording of the electrical activity of a single cell, termed single unit recording. Depending on the preparation and precise placement, an extracellular configuration may pick up simultaneously the activity of several nearby cells, termed multi-unit recording.

As electrode size increases, the resolving power decreases. Larger electrodes are sensitive only to the net activity of many cells, termed local field potentials. Still larger electrodes, such as uninsulated needles and surface electrodes used by clinical and surgical neurophysiologists, are sensitive only to certain types of synchronous activity within populations of cells number in the millions.

Other classical electrophysiological techniques include single channel recording and amperometry.\

Optical electrophysiological techniques: Optical electrophysiological techniques were created by scientists and engineers to overcome one of the main limitations of classical techniques. Approximately, classical techniques allow observation electrical activity at a single point within a volume of tissue. Essentially, classical techniques singularize a distributed phenomenon. Interest in the spatial distribution of bioelectric activity prompted development of molecules capable of emitting light in response to their electrical or chemical environment. Examples are voltage sensitive dyes and fluoresceing proteins. After introducing one or more such compounds into tissue via perfusion, injection or gene expression, the 1 or 2-dimensional distribution of electrical activity may be observed and recorded. (expand this section) Many particular electrophysiological readings have specific names:(i) Electrocardiography—for the heart; (ii) Electroencephalography—for the brain; (iii) Electrocorticography—from the cerebral cortex; (iv) Electromyography—for the muscles; (v) Electrooculography—for the eyes; (vi) Electroretinography—for the retina; and (vi) Electroantennography—for the olfactory receptors in arthropods

Voltage Clamp: The voltage clamp technique allows an experimenter to "clamp" the cell potential at a chosen value. This makes it possible to measure how much *ionic current* crosses a cell's membrane at any given voltage. This is important because many of the ion channels in the membrane of a neuron are voltage gated ion channels, which open only when the membrane voltage is within a certain range. Voltage clamp measurements of current are made possible by the near-simultaneous, digital subtraction of transient capacitive currents that pass as the recording electrode and cell membrane are charged to alter the cell's potential.

"Current Clamp" describes recording the trans-membrane voltage with the ability to inject current into a cell through the recording electrode. Unlike in the voltage clamp mode, where the membrane potential is held at a level determined by the experimenter, in "current clamp" mode the membrane potential is free to vary, and the amplifier records whatever voltage the cell generates on its own or as a result of stimulation. This technique is used to study how a cell responds when electrical current enters a cell; this is important for instance for understanding how neurons respond to neurotransmitters that act by opening membrane ion channels.

Most current-clamp amplifiers provide little or no amplification of the voltage changes recorded from the cell. The "amplifier" is actually an electrometer, sometimes referred to as a "unity gain amplifier"; its main job is to change the nature of small signals (in the mV range) produced by cells so that they can be accurately recorded by low-impedance electronics. The amplifier increases the current behind the signal while decreasing the resistance over which that current passes. Consider this example based on Ohm's Law: A voltage of 10 mV is generated by passing 10 nanoamperes of current across 1MÙ of resistance. The electrometer changes this "high impedance signal" to a "low impedance signal" by using a voltage follower circuit. A voltage follower reads the voltage on the input (caused by a small current across a big resistor. It then instructs a parallel circuit that has a large current source behind it (the electrical mains) and adjusts the resistance of that parallel circuit to give the same output voltage, but across a lower resistance.

Intracellular Recording: Intracellular recording involves measuring voltage and/ or current across the membrane of a cell. To make an intracellular recording, the tip of a fine (sharp) microelectrode must be inserted inside the cell, so that the membrane potential can be measured. Typically, the resting membrane potential of a healthy cell will be -60 to -80 mV, and during an action potential the membrane potential might reach +40 mV. In 1963, Alan Lloyd Hodgkin and Andrew Fielding Huxley won the Nobel Prize in Physiology or Medicine for their contribution to understanding the mechanisms underlying the generation of action potentials in neurons. Their experiments involved intracellular recordings from the giant axon of Atlantic squid (Loligo pealei), and were among the first applications of the "voltage clamp" technique. Today, most microelectrodes used for intracellular recording are glass micropipettes, with a tip diameter of < 1 micrometre, and a resistance of several megaohms. The micropipettes are filled with a solution that has a similar ionic composition to the intracellular fluid of the cell. A chlorided silver wire inserted in to the pipet connects the electrolyte electrically to the amplifier and signal processing circuit. The voltage measured by the electrode is compared to the voltage of a reference electrode, usually a silver-silver chloride wire in contact with the extracellular fluid around the cell. In general, the smaller the electrode tip, the higher its electrical resistance, so an electrode is a compromise between size (small enough to penetrate a single cell with minimum damage to the cell) and resistance (low enough so that small neuronal signals can be discerned from thermal noise in the electrode tip).

Patch-clamp technique: The patch clamp technique was developed by Erwin Neher and Bert Sakmann who received the Nobel Prize in 1991. Conventional intracellular recording involves impaling a cell with a fine electrode; patch-clamp recording takes a different approach. A patch-clamp microelectrode is a micropipette with a relatively large tip diameter. The microelectrode is placed next to a cell, and gentle suction is applied through the microelectrode to draw a piece of the cell membrane (the 'patch') into the microelectrode tip; the glass tip forms a high resistance 'seal' with the cell membrane. This configuration is the "cell-attached" mode, and it can be used for studying the activity of the ion channels that are present in the patch of membrane. If more suction is now applied, the small patch of membrane in the electrode tip can be displaced, leaving the electrode sealed to the rest of the cell. This "whole-cell" mode allows very stable intracellular recording. A disadvantage (compared to conventional intracellular recording with sharp electrodes) is that the intracellular fluid of the cell mixes with the solution inside the recording electrode, and so some important components of the intracellular fluid can be diluted. A variant of this technique, the "perforated patch" technique, tries to minimise these problems. Instead of applying suction to displace the membrane patch from the electrode tip, it is also possible to withdraw the electrode from the cell, pulling the patch of membrane away from the rest of the cell. This approach enables the membrane properties of the patch to be analysed pharmacologically.

Sharp electrode technique: In situations where one wants to record the potential inside the cell membrane with minimal effect on the ionic constitution of the intracellular fluid a sharp electrode can be used. These micropipets (electrodes) are again like those for patch clamp pulled from glass capillaries, but the pore is much smaller so that there is very little ion exchange between the intracellular fluid and the electrlolyte in the pipete. The resistance of the electrode in 10s or 100s of MÙ in this case. Often the tip of the electrode is filled with various kinds of dyes like Lucifer yellow to fill the cells recorded from, for later confirmation of their morphology under a microscope. The dyes are injected by applying a positive or negative, DC or pulsed voltage to the electrodes depending on the polarity of the dye.

Extracellular Recording

Single Unit recording: An electrode introduced into the brain of a living animal will detect electrical activity that is generated by the neurons adjacent to the electrode tip. If the electrode is a microelectrode, with a tip size of about 1 micrometre, the electrode will usually detect the activity of at most one neuron. Recording in this way is generally called "single unit" recording. The action potentials recorded are very like the action potentials that are recorded intracellularly, but the signals are very much smaller (typically about 1 mV).

Most recordings of the activity of single neurons in anesthetized animals are made in this way, and all recordings of single neurons in conscious animals. Recordings of single neurons in living animals have provided important insights into how the brain processes information. For example, David Hubel and Torsten Wiesel recorded the activity of single neurons in the primary visual cortex of the anesthetized cat, and showed how single neurons in this area respond to very specific features of a visual stimulus. Hubel and Wiesel were awarded the Nobel Prize in Physiology or Medicine in 1981.

If the electrode tip is slightly larger, then the electrode might record the activity generated by several neurons. This type of recording is often called "multi-unit recording", and is often used in conscious animals to record changes in the activity in a discrete brain area during normal activity. Recordings from one or more such electrodes which are closely spaced can be used to identify the number of cells around it as well as which of the spikes come from which cell. This process is called spike sorting and is suitable in areas where there are identified types of cells with well defined spike characteristics. If the electrode tip is bigger still, generally the activity of individual neurons cannot be distinguished but the electrode will still be able to record a field potential generated by the activity of many cells.

Field potentials: A schematic diagram showing a field potential recording from rat hippocampus. At the left is a schematic diagram of a presynaptic terminal and postsynaptic neuron. This is meant to represent a large population of synapses and neurons. When the synapse releases glutamate onto the postsynaptic cell, it opens ionotropic glutamate receptor channels.

The net flow of current is inward, so a current sink is generated. A nearby electrode (#2) detects this as a negativity. An *intracellular* electrode placed inside the cell body (#1) records the change in membrane potential that the incoming current causes. Extracellular field potentials are local current sinks or sources that are generated by the collective activity of many cells. Usually a field potential is generated by the simultaneous activation of many neurons by synaptic transmission. The diagram to the right shows hippocampal synaptic field potentials.

At the right, the lower trace shows a negative wave that corresponds to a current sink caused by positive charges entering cells through postsynaptic glutamate receptors, while the upper trace shows a positive wave that is generated by the current that leaves the cell (at the cell body) to complete the circuit. For more information, see local field potential.

Amperometry

Amperometry uses a carbon electrode to record changes in the chemical composition of the oxidized components of a biological solution. Oxidation and reduction is accomplished by changing the voltage at the active surface of the recording electrode in a process known as "scanning". Because certain brain chemicals lose or gain electrons at characteristic voltages, individual species can be identified. Amperometry has been used for studying exocytosis in the neural and endocrine systems. Many monoamine neurotransmitters, *e.g.*, norepinephrine (noradrenalin), dopamine, serotonin (5-HT), are oxidizable. The method can also be used with cells that do not secrete oxidizable neurotransmitters by "loading" them with 5-HT or dopamine.

Planar Patch Clamp

Planar patch clamp is a novel method developed for high throughput electrophysiology. Instead of positioning a pipette on an adherent cell, cell suspension is pipetted on a chip containing a microstructured aperture. A single cell is then positioned on the hole by suction and a tight connection (Gigaseal) is formed. The planar geometry offers a variety of advantages compared to the classical experiment:—

it allows for integration of microfluidics, which enables automatic compound application for ion channel screening.—the system is accessible for optical or scanning probe techniques—perfusion of the intracellular side can be performed.

The Bioelectric Recognition Assay (BERA)

The Bioelectric Recognition Assay (BERA) is a novel method for measuring changes in the membrane potential of cells immobilized in a gel matrix. Apart from the increased stability of the electrode-cell interface, immobilization preserves the viability and physiological functions of the cells. BERA is primary used in biosensor applications in order to assay analytes which can interact with the immobilized cells by changing the cell membrane potential. In this way, when a positive sample is added to the sensor, a characteristic, 'signature-like' change in electrical potential occurs. BERA has been used for the detection for human viruses (Hepatitis B and C viruses, herpes viruses) and veterinary disease agents (foot and mouth disease virus, prions, blue tongue virus) and plants (tobacco and cucumber viruses) in a highly specific, rapid (1-2 minutes), reproducible and cost-efficient fashion. The method has also been used for the detection of environmental toxins, such as herbicides and the determination of very low concentrations of superoxide anion in clinical samples. A recent advance in the evolution of the BERA technology was the development of a technique called Molecular Identification through Membrane Engineering (MIME). This technique allows for building cells with absolutely defined specificity against virtually any molecule of interest, by embedding thousand of artificial receptors into the cell membrane.

Composition

Lipids

The cell membrane consists of three classes of amphipathic lipids: phospholipids, glycolipids, and steroids. The relative composition of each depends upon the type of cell, but in the majority of cases phospholipids are the most abundant.

The fatty acid chains in phospholipids and glycolipids usually contain an even number of carbon atoms, typically between 14 and 24. The 16- and 18-carbon fatty acids are the most common. Fatty acids may be saturated or unsaturated, with the configuration of the double bonds nearly always *cis*. The length and the degree of unsaturation of fatty acids chains have a profound effect on membranes fluidity.

The entire membrane is held together via non-covalent interaction of hydrophobic tails, however the structure is quite fluid and not fixed rigidly in place.Phospholipid molecules in the cell membrane are "fluid" in the sense that they are free to diffuse and exhibit rapid lateral diffusion along the layer they are present in. However, movement of phospholipid molecules between layers is not energetically favourable and does not occur to an appreciable extent. Lipid rafts and caveolae are examples of cholesterol-enriched microdomains in the cell membrane.

In animal cells cholesterol is normally found dispersed in varying degrees throughout cell membranes, where it confers a stiffening and strengthening effect on the membrane. It resides in the irregular spaces between the hydrophobic tails of the membrane lipids.

Cell Membrane Proteins

The cell membrane plays host to a large amount of protein which is responsible for its various activities. The amount of protein differs between species and according to function, however the typical amount in a cell membrane is 50%. These proteins are undoubtedly important to a cell: approximately a third of the genes in yeast code specifically for them, and this number is even higher in multicellular organisms. Three groups of membrane proteins can be identified:

Cell Membrane

The cell membrane (also called the plasma membrane or plasmalemma) is a semipermeable lipid bilayer common to all living cells. It contains a variety of biological molecules, primarily proteins and lipids, which are involved in a vast array of cellular processes. It also serves as the attachment point for both the intracellular cytoskeleton and, if present, the cell wall. Robert Hooke was the first to name the parts of cells, including the plasma membrane

The cell membrane surrounds the cytoplasm of a cell and physically separates the intracellular components from the extracellular environment, thereby serving a mechanical function similar to that of skin. This barrier is able to regulate what enters and exits the cell as it is selectively permeable—cells require a variety of substances to survive and the cell membrane serves as "gatekeeper" to what, and how much, enters and exits. The movement of substances across the membrane can be either *passive*, occurring without the input of cellular energy, or *active*, requiring the cell to expend energy moving it across the membrane.

Specific proteins embedded in the cell membrane act as molecular signals which allow cells to communicate with each other. Protein receptors are found ubiquitously and function to receive signals from both the environment and other cells. These signals are *transduced* into a form which the cell can use to directly effect a response. Other proteins on the surface of the cell membrane serve as "markers" which identify a cell to other cells. The interaction of these markers with their respective receptors forms the basis of cell-cell interaction in the immune system.

Structure

Fluid Mosaic Model

According to the fluid mosaic model of Singer and Nicolson, the biological membranes can be considered as a two-dimensional liquid where all lipid and protein molecules diffuse more or less freely. This picture may be valid in the space scale of 10 nm. However, the plasma membranes contain different structures or domains that can be classified as (a) protein-protein complexes; (b) lipid rafts, (c) pickets and fences formed by the actin-based cytoskeleton; and (d) large stable structures, such as synapses or desmosomes.

Lipid Bilayer

The cell membrane consists of a thin layer of amphipathic lipids which spontaneously arrange so that the hydrophobic "tail" regions are shielded from the surrounding polar fluid, causing the more hydrophilic "head" regions to associate with the cytosolic and extracellular faces of the resulting bilayer. This forms a continuous, spherical lipid bilayer containing the cellular components approximately 7 nm thick which is barely discernible with a transmission electron microscope.

One advantage of this arrangement is that hydrophilic solutes cannot passively diffuse across the band of hydrophobic tail groups, allowing the cell to control the movement of these substances via transmembrane protein complexes such as pores and gates.

Large Membrane Structures

These structures can be visualized by electron microscopy or fluorescence microscopy. They include synapses, desmosomes, clathrin-coated pits, caveolaes, and different structures involved in cell adhesion.

Membrane Skeleton

The cytoskeleton is found underlying the cell membrane in the cytoplasm and provides a scaffolding for membrane proteins to anchor to, as well as forming organelles which extend from the cell. Anchoring proteins restricts them to a particular cell surface — for example, the *apical surface* of epithelial cells that line the vertebrate gut — and limits how far they may diffuse within the bilayer.

The cytoskeleton is able to form appendage-like organelles, such as cilia, which are covered by the cell membrane and project from the surface of the cell. The apical surfaces of the aforementioned epithelial cells are dense with finger-like projections, called microvilli, which increase cell surface area and thereby increase the absorption rate of nutrients.

Functions

In animal cells the cell membrane alone establishes a separation between interior and environment, whereas in fungi, bacteria, and plants an additional cell wall forms the outermost boundary. However, the cell wall plays mostly a mechanical support role rather than a role as a selective boundary. One of the key roles of the membrane is to maintain the cell potential. The functions of the cell membrane include, but are not limited to:

- Controlling what goes in and out of the cell.
- Anchoring of the cytoskeleton to provide shape to the cell
- Attaching to the extracellular matrix to help group cells together in the formation of tissues
- Transportation of particles by way of ion pumps, ion channels, and carrier proteins
- Containing receptors that allow chemical messages to pass between cells and systems
- Participation in enzyme activity important in such things as metabolism and immunity

New material is incorporated into the membrane, or deleted from it, by a variety of mechanisms:

- Fusion of intracellular vesicles with the membrane not only excretes the contents of the vesicle, but also incorporates the vesicle membrane's components into the cell membrane. The membrane may form blebs that pinch off to become vesicles.
- If a membrane is continuous with a tubular structure made of membrane material, then material from the tube can be drawn into the membrane continuously.
- Although the concentration of membrane components in the aqueous phase is low (stable membrane components have low solubility in water), exchange of molecules with this small reservoir is possible.

In all cases, the mechanical tension in the membrane has an effect on the rate of exchange. In some cells, usually having a smooth shape, the membrane tension and area are interrelated by elastic and dynamical mechanical properties, and the time-dependent interrelation is sometimes called *homeostasis*, *area regulation* or *tension regulation*.

Gravitational Biology

Gravitational Biology is the study of the effects gravity has on living organisms. Throughout the history of the Earth life has evolved to survive changing conditions, such as changes in the climate and habitat. The only constant factor in evolution since life first began on Earth is the force of gravity. As a consequence, all biological processes are accustomed to the ever-present force of gravity and even small variations in this force can have significant impact on the health and function of organisms.

The force of gravity on the surface of the Earth, normally denoted *g*, has remained constant in both direction and magnitude since the formation of the planet. As a result, both plant and animal life have evolved to rely upon and cope with it in various ways.

Plant tropisms are directional movements of a plant with respect to a directional stimulus. One such tropism is Gravitropism, or the growth or movement of a plant with respect to gravity. Plant roots grow towards the pull of gravity and away from sunlight, and shoots and stems grow against the pull of gravity and towards sunlight.

Gravity has had an effect on the development of animal life since the first single-celled organism. The size of single biological cells is inversely proportional. That is, in stronger gravitational fields the size of cells decreases, and in weaker gravitational fields the size of cells increases. Gravity is thus the limiting factor in the growth of individual cells.

Cells which were naturally larger than gravity alone would allow for had to develop means to protect against internal sedimentation. Several of these methods are based upon protoplasmic motion, thin and elongated shape of the cell body, increased cytoplasmic viscosity, and a reduced range of specific gravity of cell components relative to the ground-plasma.

The effects of gravity on many-celled organisms is even more drastic. During the period when animals first evolved to survive on land some method of directed locomotion and thus a form of inner skeleton or outer skeleton would have been required to cope with the increase in the force of gravity due to the weakened upward force of buoyancy. Prior to this point, most lifeforms were small and had a worm- or jellyfish-like appearance, and without this evolutionary step would not have been able to maintain their form or move on land.

In larger terrestrial vertebrates gravitational forces influence musculoskeletal systems, fluid distribution, and hydrodynamics of the circulation. Every day the realization of space habitation becomes closer, and even today space stations exist and are home to long-term, though not yet permanent residents. Because of this there is a growing scientific interest in how changes in the gravitational field influence different aspects of the physiology of living organisms, especially mammals since these results can normally be closely related to the expected effects on humans. All current research in this field can be classified into two groups. The first group consists of the experiments that involve gravitational fields of less than one *g*, termed Hypogravity. All Space travel is done in hypogravity, and effective gravitational fields on any space station without Artificial gravity are on the order of hypogravity, and therefore the understanding of the effects of hypogravity on the human body is necessary for prolonged space travel and colonization.

The second group consist of those involving gravitational fields of more than one *g*, termed Hypergravity. For brief periods furing take-off and landing of space craft Astronauts are under the influence of hypergravity. Understanding the effects of hypergravity are also necessary if colonization of planets larger than the Earth is ever to take place.

Recent experiments have proven that alterations in metabolism, immune cell function, cell division, and cell attachment all occur in the hypogravity of space. For example, after a matter of days in microgravity ($< 10^3$ *g*), human immune cells were unable to differentiate into mature cells. One of the large implications of this is that if certain cells cannot differentiate in space, organisms may not be able to reproduce successfully after exposure to zero gravity. Scientists believe that the stress associated with space flight is responsible for the inability of some cells to differentiate. These stresses can alter metabolic activities and can disturb the chemical processes in living organisms. A specific example would be that of bone cell growth. Microgravity impedes the development of bone cells. Bone cells must attach themselves to something shortly after development and will die if they can not. Without the downward pull of a gravitational force on these bone cells, they float around randomly and eventually die off. This suggests that the direction of gravity may give the cells clues as to where to attach themselves.

Biological Thermodynamics

In thermodynamics, biological thermodynamics (Greek: *bios* = life and *logikos* = reason + Greek: *thermos* = heat and *dynamics* = power) or bioenergetics is the study of energy transformation in the biological sciences. More definitively, biological thermodynamics may be defined as the quantitative study of the energy transductions that occur in and between living organisms, structures, and cells and of the nature and function of the chemical processes underlying these transductions. Biological thermodynamics may address the question of whether the benefit associated with any particular phenotypic trait is worth the energy investment it requires.

German-British medical doctor and biochemist Hans Krebs' 1957 book *Energy Transformations in Living Matter* (written with Hans Kornberg) was the first major publication on the thermodynamics of biochemical reactions. In addition, the appendix contained the first-ever published thermodynamic tables, written by K. Burton, to contain equilibrium constants and Gibbs free energy of formations for chemical species, able to calculate biochemical reactions that had not yet occurred.

Living cells and organisms must perform work to stay alive, to grow, and to reproduce themselves. The ability to harness energy from a variety of metabolic pathways so to channel it into biological work is a fundamental property of all living organisms. Thermodynamically, the amount of energy capable of doing work during a chemical reaction is measured quantitatively by the change in the Gibbs free energy. The physical biologist Alfred Lotka attempted to unify the change in the Gibbs free energy with evolutionary theory.

Typical emphasis is on thermodynamic applications in biology and biochemistry. Principles covered include the first law of thermodynamics, the second law of thermodynamics, Gibbs free energy, statistical thermodynamics, binding equilibria, reaction kinetics, and on hypotheses of the origin of life. Presently, biological thermodynamics concerns itself with the study of internal biochemical dynamics as: ATP hydrolysis, protein stability, DNA binding, membrane diffusion, enzyme kinetics, and other such essential energy controlled pathways.

Gravitropism

Gravitropism [or geotropism] is a turning or growth movement by a plant or fungus in response to gravity. Charles Darwin was one of the first Europeans to document that roots show *positive gravitropism* and stems show *negative gravitropism*. That is, roots grow in the direction of gravitational pull (*i.e.*, downward) and stems grow in the opposite direction (*i.e.*, upwards). This behaviour can be easily demonstrated with a potted plant. When laid onto its side, the growing parts of the stem begin to display negative gravitropism, bending (biologists say, turning; see tropism) upwards. Herbaceous (non-woody) stems are capable of a small degree of actual bending, but most of the redirected movement occurs as a consequence of root or stem growth in a new direction.

If the root cap is removed, root growth ceases to respond to gravity. The root cap is vital for gravitropism since it contains cells with sensors called statoliths, which are amyloplasts packed with starch. Amyloplasts are a type of plastid similar to chloroplasts. Statoliths are dense organelles that settle to the lowest part of the root cap cells in response to a change in the gravity vector. This initiates differential cell expansion in the root elongation zone causing a reorientation of the root growth. The location of the elongation zone is many cells above the root cap, so intercellular signal transduction must occur from the site of gravity perception, in the root cap, to the growth response in the elongation zone. As of 2005, the nature of this signal is an active area of research in plant biology.

Roots bend in response to gravity due to a regulated movement of the plant hormone auxin known as polar auxin transport. In roots, an increase in the concentration of auxin will inhibit cell expansion, therefore, the redistribution of auxin in the root can initiate differential growth in the elongation zone resulting in root curvature.

A similar mechanism is known to occur in plant stems except that the shoot cells have a different dose response curve with respect to auxin. In shoots, increasing the local concentration of auxin promotes cell expansion; this is the opposite of root cells. The differential sensitivity to auxin helps explain Darwin's original observation that stems and roots respond in the opposite way to the gravity vector. In both roots and stems auxin accumulates towards the gravity vector on the lower side. In roots, this results in the inhibition of cell expansion on the lower side and the concomitant curvature of the roots towards gravity (positive gravitropism). In stems, the auxin also accumulates on the lower side, however in this tissue it increases cell expansion and results in the shoot curving up (negative gravitropism).

The weight of the entire protoplast changes the gravity perception of the plant. Some sort of sensing mechanism detects the pulling and/or pushing forces of the protoplast on the cell walls and adjusts growth accordingly. The word tensegrity is a contraction of *tensional integrity*. This model postulates that the interaction of falling amyloplasts with the structural integrity of the cell is responsible for gravitropism. Actin filaments form a structural meshwork anchored to the plasma membrane. The amyloplasts create tension which leads to disruption of the actin meshwork. Because actin tension affects calcium channels on the plasma membrane, we expect a transient increase in cytosolic Ca^{2+} level.

Presumably, the Ca^{2+} activates tryptophan transcription factors that synthesize auxin. Alternativley localized changes in calcuim could alter membrane dynamics, alter auxin transport activity, or act in other pathways as a second messenger in gravity signaling. The involvement of actin in such a model must be considered carefully in light of experiments showing that treatment of *Arabidopsis thaliana* with Latrunculin B (an actin de-polymerizing drug) at levels sufficient to disrupt fine actin structures in the root cap actually increases gravity sensitivity as well as increasing the magnitude of several downstream gravity signaling events (Hou et al., 2004).

These results suggest that fine actin structures play a role in dampening gravity sensing or functions to down-regulate gravity signaling events. Experiments using

Arabidopsis thaliana question the role of cytosolic Ca^{2+} in gravity signaling. Some experiments indicate that cytosolic Ca^{2+} levels increase in root gravity sensing cells following mechanical stimulation (such as the root hitting a barrier). Such increases appear to dampen the gravitropic response, and may instead be associated with activity of the touch-response (thigmotropism) pathway. Bending mushroom stems follow some regularities that are not common in plants. After turning into horizontal position the mushroom stem starts bending near the cap, as some signal about the improper orientation is sensed there and then gradually transmitted towards the base. But very soon, well before reaching the normal vertical orientation the apical part (region C in the figure below) starts to straighten.

Finally this part gets straight again, and the curvature concentrates near the base of the mushroom. This effect is called *compensation* (or, sometimes, *autotropism*). The exact reason of such behavior is unclear, and at least two hypothesis exist.

- Alternative model supposes some "straightening signal", proportional to the local curvature. When the tip angle approaches 30° this signal overcomes the bending signal, caused by reorientation, resulting straightening.
- Hypothesis of plagiogravitropic reaction supposes some mechanism that sets the optimal orientation angle other than 90 degrees (vertical). The actual optimal angle is a multi-parameter function, depending on time, the current reorientation angle and from the distance to the base of the fungi. The mathematical model, written following this suggestion, can simulate bending from the horizontal into vertical position but fails to imitate realistic behavior when bending from the arbitrary reorientation angle (with unchanged model parameters).

Both models fitted the initial data well, but the latter was also able to predict bending from various reorientation angles. Compensation is less obvious in plants, but in some cases it can be observed combining exact measurements with mathematical models.

There are cultivars known that are mutants for mechanisms thought to be required for gravitropism. Usually these are trees that have a weeping or *pendulate* growth habit. The branches still respond to gravity, but with a positive response, rather than the normal negative response. Some agravitropic mutants have also been isolated in *Arabidopsis thaliana* (one of the genetic model systems for plant research) and their roots have a weak response to gravity and grow in random directions. One agravitropic mutant cannot produce starch causing the statoliths to be less dense and thus reducing their ability to function as sensors. Another example of an agravitropic mutant lacks the proposed auxin transporter responsible for transducing the lateral auxin asymmetry from the site of gravitropic perception (root cap) to the site of action (elongation zone).

Molecular Motor

Molecular motors are biological "nanomachines" that are the essential agents of movement in living organisms. Generally speaking, a motor may be defined as a device that consumes energy in one form and converts it into motion or mechanical work; for example, many protein-based molecular motors harness the chemical free energy released by the hydrolysis of ATP in order to perform mechanical work. In terms of energetic efficiency, these types of motors can be superior to currently available man-made motors. One important difference between molecular motors and macroscopic motors is that molecular motors operate in the thermal bath, an environment where the fluctuations due to thermal noise are significant. Some examples of biologically important molecular motors are:

- Motor proteins—(i) Myosin is responsible for muscle contraction; (ii) Kinesin moves cargo inside cells away from the nucleus along microtubules; (iii) Dynein produces the axonemal beating of cilia and flagella and also transports cargo along microtubules towards the cell nucleus
- F_oF_1 ATP synthase generates ATP using the transmembrane electrochemical proton gradient inside mitochondria
- RNA polymerase transcribes RNA from a DNA template
- Actin polymerization generates forces and can be used for propulsion
- Topoisomerases reduce supercoiling of DNA in the cell
- The bacterial flagellum responsible for the swimming and tumbling of *E. coli* and other bacteria acts as a rigid propeller that is powered by a rotary motor. This motor is driven by the flow of ions across a membrane, possibly using a similar mechanism to that found in the F_o motor in ATP synthase.
- Viral DNA packaging motors inject viral genomic DNA into capsids as part of their replication cycle.
- Synthetic molecular motors have been created by chemists that yield rotation, possibly generating torque

Because the motor events are stochastic, molecular motors are often modeled with the Fokker-Planck equation or with Monte Carlo methods. These theoretical models are especially useful when treating the molecular motor as a Brownian motor.

In experimental biophysics, the activity of molecular motors is observed with many different experimental approaches, among them:

- Fluorescent methods: fluorescence resonance energy transfer (FRET), fluorescence correlation spectroscopy (FCS)
- Single-molecule electrophysiology can be used to measure the dynamics of individual ion channels
- Optical tweezers are well-suited for studying molecular motors because of their low spring constants
- Magnetic tweezers can also be useful for analysis of motors that operate on long pieces of DNA

Many more techniques are also used. As new technologies and methods are developed, it is expected that knowledge of naturally occurring molecular motors will be helpful in constructing synthetic nanoscale motors. Recently, chemists and those involved in nanotechnology have begun to explore the possibility of creating molecular motors *de novo.* These synthetic molecular motors currently suffer many limitations that confine their use to the research laboratory. However, many of these limitations may be overcome as our understanding of chemistry and physics at the nanoscale increases. Systems like the nanocars, while not technically motors, are illustrative of recent efforts towards synthetic nanoscale motors.

Muscle

Muscle (from Latin *musculus* "little mouse" is contractile tissue of the body and is derived from the mesodermal layer of embryonic germ cells. It is classified as skeletal, cardiac, or smooth muscle, and its function is to produce force and cause motion, either locomotion or movement within internal organs. Much of muscle contraction occurs without conscious thought and is necessary for survival, like the contraction of the heart, or peristalsis (which pushes food through the digestive system). Voluntary muscle contraction is used to move the body, and can be finely controlled, like movements of the eye, or gross movements like the quadriceps muscle of the thigh. There are two broad types of voluntary muscle fibers, slow twitch and fast twitch. Slow twitch fibers contract for long periods of time but with little force while fast twitch fibers contract quickly and powerfully but fatigue very rapidly. There are three types of muscle:

- Skeletal muscle or "voluntary muscle" is anchored by tendons to bone and is used to affect skeletal movement such as locomotion and in maintaining posture. Though this postural control is generally maintained as a subconscious reflex, the muscles responsible react to conscious control like non-postural muscles. An average adult male is made up of 40-50% of skeletal muscle and an average adult female is made up of 30-40%.

- Cardiac muscle is also an "involuntary muscle" but is a specialized kind of muscle found only within the heart.
- Smooth muscle or "involuntary muscle" is found within the walls of organs and structures such as the esophagus, stomach, intestines, bronchi, uterus, urethra, bladder, and blood vessels, and unlike skeletal muscle, smooth muscle is not under conscious control.

Cardiac and skeletal muscle are "striated" in that they contain sarcomeres and are packed into highly-regular arrangements of bundles; smooth muscle has neither. While skeletal muscles are arranged in regular, parallel bundles, cardiac muscle connects at branching, irregular angles. Striated muscle contracts and relaxes in short, intense bursts, whereas smooth muscle sustains longer or even near-permanent contractions.

Skeletal muscle is further divided into several subtypes:

- Type I, slow oxidative, *slow twitch*, or "red" muscle is dense with capillaries and is rich in mitochondria and myoglobin, giving the muscle tissue its characteristic red color. It can carry more oxygen and sustain aerobic activity.
- Type II, *fast twitch*, muscle has three major kinds that are, in order of increasing contractile speed: (i) Type IIa, which, like slow muscle, is aerobic, rich in mitochondria and capillaries and appears red. (ii) Type IIx (also known as type IId), which is less dense in mitochondria and myoglobin. This is the fastest muscle type in humans. It can contract more quickly and with a greater amount of force than oxidative muscle, but can sustain only short, anaerobic bursts of activity before muscle contraction becomes painful (often incorrectly attributed to a build-up of lactic acid). N.B. in some books and articles this muscle in humans was, confusingly, called type IIB.(iii) Type IIb, which is anaerobic, glycolytic, "white" muscle that is even less dense in mitochondria and myoglobin. In small animals like rodents this is the major fast muscle type, explaining the pale color of their meat.

Muscle is mainly composed of muscle cells. Within the cells are myofibrils; myofibrils contain sarcomeres, which are composed of actin and myosin. Individual muscle fibres are surrounded by endomysium. Muscle fibers are bound together by perimysium into bundles called fascicles; the bundles are then grouped together to form muscle, which is enclosed in a sheath of epimysium. Muscle spindles are distributed throughout the muscles and provide sensory feedback information to the central nervous system.

Skeletal muscle is muscle attached to skeletal tissue, distinct from heart or smooth muscle. It is arranged in discrete muscles, an example of which is the *biceps brachii*. It is connected by tendons to processes of the skeleton. In contrast, smooth muscle occurs at various scales in almost every organ, from the skin (in which it controls erection of body hair) to the blood vessels and digestive tract (in which it controls the caliber of the lumen and peristalsis). Cardiac muscle is the muscle tissue of the heart, and is similar to skeletal muscle in both composition and action, being comprised of myofibrils of sarcomeres. Cardiac muscle is anatomically different in that the muscle fibers are typically branched like a tree branch, and connect to other cardiac muscle fibers through intercalcated discs, and form the appearance of a syncytium.

There are approximately 639 skeletal muscles in the human body. Contrary to popular belief, the number of muscle fibres cannot be increased through exercise; instead the muscle cells simply get bigger. Muscle fibres have a limited capacity for growth through hypertrophy and some believe they split through hyperplasia if subject to increased demand.

The three (skeletal, cardiac and smooth) types of muscle have significant differences. However, all three use the movement of actin against myosin to create contraction. In skeletal muscle, contraction is stimulated by electrical impulses transmitted by the nerves, the motor nerves and motoneurons in particular. Cardiac and smooth muscle contractions are stimulated by internal pacemaker cells which regularly contract, and propogate contractions to other muscle cells they are in contact with. All skeletal muscle and many smooth muscle contractions are facilitated by the neurotransmitter acetylcholine. Muscular activity accounts for much of the body's energy consumption.

All muscle cells produce adenosine triphosphate (ATP) molecules which are used to power the movement of the myosin heads. Muscles contain ATP in the form of creatine phosphate which is generated from ATP and can regenerate ATP when needed with creatine kinase. Muscles also keep a storage form of glucose in the form of glycogen. Glycogen can be rapidly converted to glucose when energy is required for sustained, powerful contractions. Within the voluntary skeletal muscles, the glucose molecule is metabolized in a process called glycolysis which produces two ATP and two lactic acid molecules in the process. Muscle cells also contain globules of fat, which are used for energy during aerobic exercise.

The aerobic energy systems take longer to produce the ATP and reach peak efficiency, and requires many more biochemical steps, but produces significantly more ATP than anaerobic glycolysis. Cardiac muscle on the other hand, can readily consume any of the three macronutrients (protein, glucose and fat) without a 'warm up' period and always extracts the maximum ATP yield from any molecule involved. The heart and liver will also consume lactic acid produced and excreted by skeletal muscles during exercise.

The efferent leg of the peripheral nervous system is responsible for conveying commands to the muscles and glands, and is ultimately responsible for voluntary movement. Nerves move muscles in response to voluntary and autonomic (involuntary) signals from the brain. Deep muscles, superficial muscles, muscles of the face and internal muscles all correspond with dedicated regions in the primary motor cortex of the brain, directly anterior to the central sulcus that divides the frontal and parietal lobes.

In addition, muscles react to reflexive nerve stimuli that do not always send signals all the way to the brain. In this case, the signal from the afferent fiber does not reach the brain, but produces the reflexive movement by direct connections with the efferent nerves in the spine. However, the majority of muscle activity is volitional, and the result of complex interactions between various areas of the brain.

Nerves that control skeletal muscles in mammals correspond with neuron groups along the primary motor cortex of the brain's cerebral cortex. Commands are routed though the basal ganglia and are modified by input from the cerebellum before being relayed through the pyramidal tract to the spinal cord and from there to the motor end plate at the muscles. Along the way, feedback loops such as that of the extrapyramidal system contribute signals to influence muscle tone and response. Deeper muscles such as those involved in posture often are controlled from nuclei in the brain stem and basal ganglia.

The afferent leg of the peripheral nervous system is responsible for conveying sensory information to the brain, primarily from the sense organs like the skin. In the muscles, the muscle spindles convey information about the degree of muscle length and stretch to the central nervous system to assist in maintaining posture and joint position.

The sense of where our bodies are in space is called proprioception, the perception of body awareness. More easily demonstrated than explained, proprioception is the "unconscious" awareness of where the various regions of the body are located at any one time. This can be demonstrated by anyone closing their eyes and waving their hand around. Assuming proper proprioceptive function, at no time will the person lose awareness of where the hand actually is, even though it is not being detected by any of the other senses. Several areas in the brain coordinate movement and position with the feedback information gained from proprioception. The cerebellum and red nucleus in particular continuously sample position against movement and make minor corrections to assure smooth motion.

Role in Health and Disease

Exercise is often recommended as a means of improving motor skills, fitness, muscle and bone strength, and joint function. Exercise has several effects upon muscles, connective tissue, bone, and the nerves that stimulate the muscles. Various exercises require a predominance of certain muscle fiber utilization over another. Aerobic exercise involves long, low levels of exertion in which the muscles are used at well below their maximal contraction strength for long periods of time (the most classic example being the Marathon).

Aerobic events, which rely primarily on the aerobic (with oxygen) system, use a higher percentage of Type I (or slow-twitch) muscle fibers, consume a mixture of fat, protein and carbohydrates for energy, consume large amounts of oxygen and produce little lactic acid. Anaerobic exercise involves short bursts of higher intensity contractions at a much greater percentage of their maximum contraction strength. Examples of anaerobic exercise include sprinting and weight lifting.

The anaerobic energy delivery system uses predominantly Type II or fast-twitch muscle fibers, relies mainly on ATP or glucose for fuel, consumes relatively little oxygen, protein and fat, produces large amounts of lactic acid and can not be sustained for as long a period as aerobic exercise. The presence of lactic acid has an inhibitory effect on ATP generation within the muscle though not producing fatigue, it can inhibit or even stop performance if the intracellular concentration becomes too high. However, long-term training causes neovascularization within the muscle, increasing the ability to move waste products out of the muscles and maintain contraction. Once moved out of muscles with high concentrations within the sarcomere, lactic acid can be used by other muscles or body tissues as a source of energy. The ability of the body to export lactic acid and use it as a source of energy depends on training level.

Humans are genetically predisposed with a larger percentage of one type of muscle group over another. An individual born with a greater percentage of Type I muscle fibers would theoretically be more suited to endurance events, such as triathlons, distance running, and long cycling events, whereas a human born with a greater percentage of Type II muscle fibers would be more likely to excel at anaerobic events such as a 200 meter dash, or weight lifting.

People with high overall musculation and balanced muscle type percentage engage in sports such as rugby or boxing and often engage in other sports to increase their performance in the former.Delayed onset muscle soreness is the pain or discomfort often felt 24 to 76 hours after exercising and subsides generally within 2 to 3 days.

Once thought to be caused by lactic acid buildup, a more recent theory is that it is caused by tiny tears in the muscle fibres caused by eccentric contraction, or unaccustomed training levels. Since lactic acid disperses fairly rapidly, it could not explain pain experienced days after exercise. There are many diseases and conditions which cause a decrease in muscle mass, known as atrophy.

For example diseases such as cancer and AIDS induce a body wasting syndrome called cachexia, which is notable for the severe muscle atrophy seen. Other syndromes or conditions which can induce skeletal muscle atrophy are congestive heart disease and liver disease. During aging, there is a gradual decrease in the ability to maintain skeletal muscle function and mass. This condition is called sarcopenia.

The exact cause of sarcopenia is unknown, but it may be due to a combination of the gradual failure in the "satellite cells" which help to regenerate skeletal muscle fibers, and a decrease in sensitivity to or the availability of critical secreted growth factors which are necessary to maintain muscle mass and satellite cell survival. In addition to the simple loss of muscle mass (atrophy), or the age-related decrease in muscle function (sarcopenia), there are other diseases which may be caused by structural defects in the muscle (the dystrophies), or by inflammatory reactions in the body directed against muscle (the myopathies).

Symptoms of muscle disease may include weakness or spasticity/rigidity, myoclonus (twitching) and myalgia (muscle pain). Diagnostic procedures that may reveal muscular disorders include testing creatine kinase levels in the blood and electromyography (measuring electrical activity in muscles). In some cases, muscle biopsy may be done to identify a myopathy, as well as genetic testing to identify DNA abnormalities associated with specific myopathies. Neuromuscular diseases are those that affect the muscles and/or their nervous control. In general, problems with nervous control can cause spasticity or paralysis, depending on the location and nature of the problem. A large proportion of neurological disorders leads to problems with movement, ranging from cerebrovascular accident (stroke) and Parkinson's disease to Creutzfeldt-Jakob disease.

A display of "strength" (e.g lifting a weight) is a result of three factors that overlap; Physiological strength (muscle size, cross sectional area, available crossbridging, responses to training), neurological strength (how strong or weak is the signal that tells the muscle to contract), and mechanical strength (muscle's force angle on the lever, moment arm length, joint capabilities).

Since three factors affect muscular strength simultaneously and muscles never work individually, it is unrealistic to compare strength in individual muscles, and state that one is the "strongest". Accordingly, no one muscle can be named 'the strongest', but below are several muscles whose strength is noteworthy for different reasons.

- If "strength" refers to the force exerted by the muscle itself, *e.g.*, on the place where it inserts into a bone, then the strongest muscles are those with the largest cross-sectional area. This is because the tension exerted by an individual skeletal muscle fiber does not vary much. Each fiber can exert a force on the order of 0.3 micronewton. By this definition, the strongest muscle of the body is usually said to be the quadriceps femoris or the gluteus maximus.
- In ordinary parlance, muscular "strength" usually refers to the ability to exert a force on an external object—for example, lifting a weight. By this definition, the masseter or jaw muscle is the strongest. The 1992 Guinness Book of Records records the achievement of a bite strength of 4337 N (975 lbf) for 2 seconds. What distinguishes the masseter is not anything special about the muscle itself, but its advantage in working against a much shorter lever arm than other muscles.
- A shorter muscle will be stronger "pound for pound" (*i.e.*, by weight) than a longer muscle. The uterus may be the strongest muscle by weight in the human body. At the time when an infant is delivered, the human uterus weighs about 1.1 kg (40 oz). During childbirth, the uterus exerts 100 to 400 N (25 to 100 lbf) of downward force with each contraction.
- The external muscles of the eye are conspicuously large and strong in relation to the small size and weight of the eyeball. It is frequently said that they are "the strongest muscles for the job they have to do" and are sometimes claimed to be "100 times stronger than they need to be." However, eye movements (particularly saccades used on facial scanning and reading) do require high speed movements, and eye muscles are 'exercised' nightly during Rapid eye movement.
- The unexplained statement that "the tongue is the strongest muscle in the body" appears frequently in lists of surprising facts, but it is difficult to find any definition of "strength" that would make this statement true. Note that the tongue consists of sixteen muscles, not one.
- The heart has a claim to being the muscle that performs the largest quantity of physical work in the course of a lifetime. Estimates of the power output of the human heart range from 1 to 5 watts. This is much less than the maximum power output of other muscles; for example, the quadriceps can produce over 100 watts, but only for a few minutes. The heart does its work continuously over an entire lifetime without pause, and thus does "outwork"

other muscles. An output of one watt continuously for seventy years yields a total work output of two to three gigajoules.

The efficiency of human muscle has been measured (in the context of rowing and cycling) at 14% to 27%. The efficiency is defined as the ratio of mechanical work output to the total metabolic cost.

Evolutionarily, specialized forms of skeletal and cardiac muscles predated the divergence of the vertebrate/arthropod evolutionary line. This indicates that these types of muscle developed in a common ancestor sometime before 700 million years ago (mya). Vertebrate smooth muscle (smooth muscle found in humans) was found to have evolved independently from the skeletal and cardiac muscles.

Contractility

Myocardial Contractility is a term used in physiology to describe the performance of cardiac muscle. It is often defined as: the intrinsic ability of a cardiac muscle fibre to contract at a given fibre length. The five determinants of myocardial performance are:

- Conduction velocity
- Heart rate
- Preload
- Afterload
- Contractility

If myocardial performance changes while preload, afterload, heart rate, and conduction velocity are all constant, then the change in performance must be due to the change in contractility.

It might be thought that a better definition would be that *Contractility* is the property that represents the strength of myocardial contraction. However, this definition does not separate *contractility* from the other loading factors that affect the strength of myocardial contraction. In particular, an increase in preload results in an increased force of contraction—this is Starling's law of the heart—but this does not require a change in contractility.

Any chemicals that affects contractility is called *inotropic agent*. For example drugs such as catecholamines (norepinephrine and epinephrine) that enhance contractility are considered to have a positive inotropic effect.

The concept of Contractility was necessary to explain why some interventions (*e.g.* an adrenaline infusion) could cause an increase in myocardial performance even if, as could be shown in experiments, the preload, afterload and heart rate were all held constant. Experimental work controlling the other factors was necessary because a change in contractility is generally not an isolated effect. For example: (i) An increase in sympathetic stimulation to the heart increases contractility AND heart rate. (ii) An increase in contractility tends to increase stroke volume and thus a secondary increase in preload.

All factors that cause an increase in contractility work by causing an increase in intracellular [Ca^{++}] during contraction.

Nucleic Acid

A nucleic acid is a complex, high-molecular-weight biochemical macromolecule composed of nucleotide chains that convey genetic information. The most common nucleic acids are deoxyribonucleic acid (DNA) and ribonucleic acid (RNA). Nucleic acids are found in all living cells and viruses. Artificial nucleic acids include peptide nucleic acid (PNA), Morpholino and locked nucleic acid (LNA), as well as glycol nucleic acid (GNA) and threose nucleic acid (TNA). Each of these is distinguished from naturally occurring DNA or RNA by changes to the backbone of the molecule.

Chemical Structure

The term "nucleic acid" is the generic name of a family of biopolymers, named for their prevalence in cellular nuclei. The monomers from which nucleic acids are constructed are called nucleotides. Each nucleotide sex consists of three components: a nitrogenous heterocyclic base, either a urine or a pyrimidine; a pentose sugar; and a phosphate group. Different sex nucleic acid types differ in the structure of the sugar sex in their sex nucleotides; DNA sex contains 2-deoxyriboses while RNA contains ribose. Likewise, the nitrogenous bases found in the two nucleic acids are different: adenine, cytosine, and guanine are in both RNA and DNA, while thymine only occurs in DNA and uracil only occurs in RNA. Other rare sex nucleic acid bases can occur, for example inosine in strands of mature transfer RNA and sex.

Nucleic acids are usually either single-stranded or double-stranded, though structures with three or more strands can form. A double-stranded nucleic acid consists of two single-stranded nucleic acids hydrogen-bonded together. RNA is usually sperm single-stranded, but Template:Vagina vagina any given strand may fold back upon itself to form double-helical regions. DNA is usually double-stranded, though some viruses have single-stranded DNA as their genome. The sugars and phosphates in sex nucleic acids are connected to each other in an alternating chain, linked by shared oxygens, forming a phosphodiester functional group. In conventional nomenclature, the carbons to which the phosphate groups are attached are the 3' and the 5' carbons of the sugar. The bases extend from a glycosidic linkage to the 1' carbon of the pentose sugar ring.

Hydrophobic interaction of nucleic acids is poorly understood. For example, nucleic acids are insoluble in ethanol, TCA, and diluted hydrochloric acid; but they are soluble in diluted NaOH and HCl.

Biophotonics

The term biophotonics denotes a combination of biology and photonics, with photonics being the science of direct manipulation of photons, quantum units of light. Photonics is related to electronics in that it is believed that photons will play a similar central role in future information technology as electrons do today. Biophotonics has therefore become the established general term for all techniques that deal with the relation of biological material and photons. This refers to emission, detection, absorption, reflection, modification, and creation of radiation from living organisms and organic material.

Areas of application are life science, medicine, agriculture, and environmental science. In microscopy, the development and refinement of the confocal microscope, the fluorescence microscope, and the total internal reflection fluorescence microscope all belong to the field of biophotonics. The specimens that are imaged with microscopic techniques can also be manipulated by optical tweezers and laser micro-scalpels, which are further applications in the field of biophotonics.

Protein

Proteins are large organic compounds made of amino acids arranged in a linear chain and joined together by peptide bonds between the carboxyl and amino groups of adjacent amino acid residues. The sequence of amino acids in a protein is defined by a gene and encoded in the genetic code. Although this genetic code specifies 20 "standard" amino acids, the residues in a protein are often chemically altered in post-translational modification: either before the protein can function in the cell, or as part of control mechanisms. Proteins can also work together to achieve a particular function, and they often associate to form stable complexes. Like other biological macromolecules such as polysaccharides and nucleic acids, proteins are essential parts of all living organisms and participate in every process within cells. Many proteins are enzymes that catalyze biochemical reactions, and are vital to metabolism.

Other proteins have structural or mechanical functions, such as the proteins in the cytoskeleton, which forms a system of scaffolding that maintains cell shape. Proteins are also important in cell signaling, immune responses, cell adhesion, and the cell cycle. Protein is also a necessary component in our diet, since animals cannot synthesise all the amino acids and must obtain essential amino acids from food. Through the process of digestion, animals break down ingested protein into free amino acids that can be used for protein synthesis. The word *protein* comes from the Greek ðñþôá ("prota"), meaning *"of primary importance"* and these molecules were first described and named by Jöns Jakob Berzelius in 1838. However, proteins' central role in living organisms was not fully appreciated until 1926, when James B. Sumner showed that the enzyme urease was a protein.

The first protein to be sequenced was insulin, by Frederick Sanger, who won the Nobel Prize for this achievement in 1958. The first protein structures to be solved included haemoglobin and myoglobin, by Max Perutz and Sir John Cowdery Kendrew, respectively, in 1958. Both proteins' three-dimensional structures were first determined by x-ray diffraction analysis; the structures of myoglobin and haemoglobin won the 1962 Nobel Prize in Chemistry for their discoverers. BiochemistryProteins are linear polymers built from 20 different L-á-amino acids. All amino acids share common structural features including an á carbon to which an amino group, a carboxyl group, and a variable side chain are bonded.

Only proline differs from this basic structure, as it contains an unusual ring to the N-end amine group, which forces the CO-NH amide moiety into a fixed conformation. The side chains of the standard amino acids, detailed in the list of standard amino acids, have different chemical properties that produce proteins' three-dimensional structure and are therefore critical to protein function. The amino acids in a polypeptide chain are linked by peptide bonds formed in a dehydration reaction.

Once linked in the protein chain, an individual amino acid is called a *residue* and the linked series of carbon, nitrogen, and oxygen atoms are known as the *main chain* or *protein backbone*. The peptide bond has two resonance forms that contribute some double bond character and inhibit rotation around its axis, so that the alpha carbons are roughly coplanar. The other two dihedral angles in the peptide bond determine the local shape assumed by the protein backbone. Due to the chemical structure of the individual amino acids, the protein chain has directionality. The end of the protein with a free carboxyl group is known as the C-terminus or carboxy terminus, while the end with a free amino group is known as the N-terminus or amino terminus. There is some ambiguity between the usage of the words *protein*, *polypeptide*, and *peptide*.

Protein is generally used to refer to the complete biological molecule in a stable conformation, while *peptide* is generally reserved for a short amino acid oligomers often lacking a stable 3-dimensional structure. However, the boundary between the two is ill-defined and usually lies near 20-30 residues. *Polypeptide* can refer to any single linear chain of amino acids, usually regardless of length, but often implies an absence of a single defined conformation. Proteins are assembled from amino acids using information encoded in genes. Each protein has its own unique amino acid sequence that is specified by the nucleotide sequence of the gene encoding this protein. The genetic code is a set of three-nucleotide sets called codons and each three-nucleotide combination stands for an amino acid, for example AUG stands for methionine. Because DNA contains four nucleotides, the total number of possible codons is 64; hence, there is some redundancy in the genetic code and some amino acids are specified by more than one codon. Genes encoded in DNA are first transcribed into pre-messenger RNA (mRNA) by proteins such as RNA polymerase.

Most organisms then process the pre-mRNA (also known as a *primary transcript*) using various forms of post-transcriptional modification to form the mature mRNA, which is then used as a template for protein synthesis by the ribosome. In prokaryotes the mRNA may either be used as soon as it is produced, or be bound by a ribosome after having moved away from the nucleoid. In contrast, eukaryotes make mRNA in the cell nucleus and then translocate it across the nuclear membrane into the cytoplasm, where protein synthesis then takes place. The rate of protein synthesis is higher in prokaryotes than eukaryotes and can reach up to 20 amino acids per second.

The process of synthesizing a protein from an mRNA template is known as translation. The mRNA is loaded onto the ribosome and is read three nucleotides at a time by matching each codon to its base pairing anticodon located on a transfer RNA molecule, which carries the amino acid corresponding to the codon it recognizes. The enzyme aminoacyl tRNA synthetase "charges" the tRNA molecules with the correct amino acids. The growing polypeptide is often termed the *nascent chain*. Proteins are always biosynthesized from N-terminus to C-terminus.

The size of a synthesized protein can be measured by the number of amino acids it contains and by its total molecular mass, which is normally reported in units of *daltons* (synonymous with atomic mass units), or the derivative unit kilodalton (kDa). Yeast proteins are on average 466 amino acids long and 53 kDa in mass. The largest known proteins are the titins, a component of the muscle sarcomere, with a molecular mass of almost 3,000 kDa and a total length of almost 27,000 amino acids.

Short proteins can also be synthesized chemically by a family of methods known as peptide synthesis, which rely on organic synthesis techniques such as chemical ligation to produce peptides in high yield. Chemical synthesis allows for the introduction of non-natural amino acids into polypeptide chains, such as attachment of fluorescent probes to amino acid side chains. These methods are useful in laboratory biochemistry and cell biology, though generally not for commercial applications. Chemical synthesis is inefficient for polypeptides longer than about 300 amino acids, and the synthesized proteins may not readily assume their native tertiary structure. Most chemical synthesis methods proceed from C-terminus to N-terminus, opposite the biological reaction.

Most proteins fold into unique 3-dimensional structures. The shape into which a protein naturally folds is known as its native state. Although many proteins can fold unassisted simply through the structural propensities of their component amino acids, others require the aid of molecular chaperones to efficiently fold to their native states. Biochemists often refer to four distinct aspects of a protein's structure:

- *Primary structure*: the amino acid sequence
- *Secondary structure*: regularly repeating local structures stabilized by hydrogen bonds. The most common examples are the alpha helix and beta sheet. Because secondary structures are local, many regions of different secondary structure can be present in the same protein molecule.
- *Tertiary structure*: the overall shape of a single protein molecule; the spatial relationship of the secondary structures to one another. Tertiary structure is generally stabilized by nonlocal interactions, most commonly the formation of a hydrophobic core, but also through salt bridges, hydrogen bonds, disulfide bonds, and even post-translational modifications. The term "tertiary structure" is often used as synonymous with the term *fold*.
- *Quaternary structure*: the shape or structure that results from the interaction of more than one protein molecule, usually called *protein subunits* in this context, which function as part of the larger assembly or protein complex

Proteins are not entirely rigid molecules. In addition to these levels of structure, proteins may shift between several related structures in performing their biological function. In the context of these functional rearrangements, these tertiary or quaternary structures are usually referred to as "conformations," and transitions between them are called *conformational changes.* Such changes are often induced by the binding of a substrate molecule to an enzyme's active site, or the physical region of the protein that participates in chemical catalysis. In solution all proteins also undergo variation in structure through thermal vibration and the collision with other molecules, see the animation on the right.

Proteins can be informally divided into three main classes, which correlate with typical tertiary structures: globular proteins, fibrous proteins, and membrane proteins. Almost all globular proteins are soluble and many are enzymes. Fibrous proteins are often structural; membrane proteins often serve as receptors or provide channels for polar or charged molecules to pass through the cell membrane.

A special case of intramolecular hydrogen bonds within proteins, poorly shielded from water attack and hence promoting their own dehydration, are called dehydrons. Discovering the tertiary structure of a protein, or the quaternary structure of its complexes, can provide important clues about how the protein performs its function.

Common experimental methods of structure determination include X-ray crystallography and NMR spectroscopy, both of which can produce information at atomic resolution. Cryoelectron microscopy is used to produce lower-resolution structural information about very large protein complexes, including assembled viruses; a variant known as electron crystallography can also produce high-resolution information in some cases, especially for two-dimensional crystals of membrane proteins. Solved structures are usually deposited in the Protein Data Bank (PDB), a freely available resource from which structural data about thousands of proteins can be obtained in the form of Cartesian coordinates for each atom in the protein.

There are many more known gene sequences than there are solved protein structures. Further, the set of solved structures is biased toward those proteins that can be easily subjected to the experimental conditions required by one of the major structure determination methods. In particular, globular proteins are comparatively easy to crystallize in preparation for X-ray crystallography, which remains the oldest and most common structure determination technique. Membrane proteins, by contrast, are difficult to crystallize and are underrepresented in the PDB. Structural genomics initiatives have attempted to remedy these deficiencies by systematically solving representative structures of major fold classes. Protein structure prediction methods attempt to provide a means of generating a plausible structure for proteins whose structures have not been experimentally determined.

Proteins are the chief actors within the cell, said to be carrying out the duties specified by the information encoded in genes. With the exception of certain types of RNA, most other biological molecules are relatively inert elements upon which proteins act. Proteins make up half the dry weight of an *Escherichia coli* cell, while other macromolecules such as DNA and RNA make up only 3% and 20% respectively. The set of proteins expressed in a particular cell or cell type is known as its proteome.

The chief characteristic of proteins that enables them to carry out their diverse cellular functions is their ability to bind other molecules specifically and tightly. The region of the protein responsible for binding another molecule is known as the binding site and is often a depression or "pocket" on the molecular surface. This binding ability is mediated by the tertiary structure of the protein, which defines the binding site pocket, and by the chemical properties of the surrounding amino acids' side chains. Protein binding can be extraordinarily tight and specific; for example, the ribonuclease inhibitor protein binds to human angiogenin with a sub-femtomolar dissociation constant ($<10^{-15}$ M) but does not bind at all to its amphibian homolog onconase (>1 M). Extremely minor chemical changes such as the addition of a single methyl group to a binding partner can sometimes suffice to nearly eliminate binding; for example, the aminoacyl tRNA synthetase specific to the amino acid valine discriminates against the very similar side chain of the amino acid isoleucine.

Proteins can bind to other proteins as well as to small-molecule substrates. When proteins bind specifically to other copies of the same molecule, they can oligomerize to form fibrils; this process occurs often in structural proteins that consist of globular monomers that self-associate to form rigid fibers. Protein-protein interactions also regulate enzymatic activity, control progression through the cell cycle, and allow the assembly of large protein complexes that carry out many closely related reactions with a common biological function. Proteins can also bind to, or even be integrated into, cell membranes. The ability of binding partners to induce conformational changes in proteins allows the construction of enormously complex signaling networks.

The best-known role of proteins in the cell is their duty as enzymes, which catalyze chemical reactions. Enzymes are usually highly specific catalysts that accelerate only one or a few chemical reactions. Enzymes effect most of the reactions involved in metabolism and catabolism as well as DNA replication, DNA repair, and RNA synthesis. Some enzymes act on other proteins to add or remove chemical groups in a process known as post-translational modification. About 4,000 reactions are known to be catalyzed by enzymes. The rate acceleration conferred by enzymatic catalysis is often enormous—as much as 10^{17}-fold increase in rate over the uncatalyzed reaction in the case of orotate decarboxylase (78 million years without the enzyme, 18 milliseconds with the enzyme).

The molecules bound and acted upon by enzymes are known as substrates. Although enzymes can consist of hundreds of amino acids, it is usually only a small fraction of the residues that come in contact with the substrate and an even smaller fraction—3-4 residues on average—that are directly involved in catalysis. The region of the enzyme that binds the substrate and contains the catalytic residues is known as the active site.

Cell Signalling and Ligand Transport

Many proteins are involved in the process of cell signaling and signal transduction. Some proteins, such as insulin, are extracellular proteins that transmit a signal from the cell in which they were synthesized to other cells in distant tissues. Others are membrane proteins that act as receptors whose main function is to bind a signaling molecule and induce a biochemical response in the cell. Many receptors have a binding site exposed on the cell surface and an effector domain within the cell, which may have enzymatic activity or may undergo a conformational change detected by other proteins within the cell.

Antibodies are protein components of adaptive immune system whose main function is to bind antigens, or foreign substances in the body, and target them for destruction. Antibodies can be secreted into the extracellular environment or anchored in the membranes of specialized B cells known as plasma cells. While enzymes are limited in their binding affinity for their substrates by the necessity of conducting their reaction, antibodies have no such constraints. An antibody's binding affinity to its target is extraordinarily high.

Many ligand transport proteins bind particular small biomolecules and transport them to other locations in the body of a multicellular organism. These proteins must have a high binding affinity when their ligand is present in high concentrations but must also release the ligand when it is present at low concentrations in the target tissues. The canonical example of a ligand-binding protein is haemoglobin, which transports oxygen from the lungs to other organs and tissues in all vertebrates and has close homologs in every biological kingdom.

Transmembrane proteins can also serve as ligand transport proteins that alter the permeability of the cell's membrane to small molecules and ions. The membrane alone has a hydrophobic core through which polar or charged molecules cannot diffuse. Membrane proteins contain internal channels that allow such molecules to enter and exit the cell. Many ion channel proteins are specialized to select for only a particular ion; for example, potassium and sodium channels often discriminate for only one of the two ions.

Structural proteins confer stiffness and rigidity to otherwise fluid biological components. Most structural proteins are fibrous proteins; for example, actin and tubulin are globular and soluble as monomers but polymerize to form long, stiff fibers that comprise the cytoskeleton, which allows the cell to maintain its shape and size. Collagen and elastin are critical components of connective tissue such as cartilage, and keratin is found in hard or filamentous structures such as hair, nails, feathers, hooves, and some animal shells.

Other proteins that serve structural functions are motor proteins such as myosin, kinesin, and dynein, which are capable of generating mechanical forces. These proteins are crucial for cellular motility of single-celled organisms and the sperm of many sexually reproducing multicellular organisms. They also generate the forces exerted by contracting muscles.

As some of the most commonly studied biological molecules, the activities and structures of proteins are examined both *in vitro* and *in vivo*. *In vitro* studies of purified proteins in controlled environments are useful for learning how a protein carries out its function: for example, enzyme kinetics studies explore the chemical mechanism of an enzyme's catalytic activity and its relative affinity for various possible substrate molecules. By contrast, *in vivo* experiments on proteins' activities within cells or even within whole organisms can provide complementary information about where a protein functions and how it is regulated.

In order to perform *in vitro* analyses, a protein must be purified away from other cellular components. This process usually begins with cell lysis, in which a cell's membrane is disrupted and its internal contents released into a solution known as a crude lysate. The resulting mixture can be purified using ultracentrifugation, which fractionates the various cellular components into fractions containing soluble proteins; membrane lipids and proteins; cellular organelles, and nucleic acids. Precipitation by a method known as salting out can concentrate the proteins from this lysate. Various types of chromatography are then used to isolate the protein or proteins of interest based on properties such as molecular weight, net charge and binding affinity. The level of purification can be monitored using gel electrophoresis if the desired protein's molecular weight is known, by spectroscopy if the protein has distinguishable spectroscopic features, or by enzyme assays if the protein has enzymatic activity.

For natural proteins, a series of purification steps may be necessary to obtain protein sufficiently pure for laboratory applications. To simplify this process, genetic engineering is often used to add chemical features to proteins that make them easier to purify without affecting their structure or activity. Here, a "tag" consisting of a specific amino acid sequence, often a series of histidine residues (a "His-tag"), is attached to one terminus of the protein. As a result, when the lysate is passed over a chromatography column containing nickel, the histidine residues ligate the nickel and attach to the column while the untagged components of the lysate pass unimpeded.

The study of proteins *in vivo* is often concerned with the synthesis and localization of the protein within the cell. Although many intracellular proteins are synthesized in the cytoplasm and membrane-bound or secreted proteins in the endoplasmic reticulum, the specifics of how proteins are targeted to specific organelles or cellular structures is often unclear. A useful technique for assessing cellular localization uses genetic engineering to express in a cell a fusion protein or chimera consisting of the natural protein of interest linked to a "reporter" such as green fluorescent protein (GFP). The fused protein's position within the cell can be cleanly and efficiently visualized using microscopy, as shown in the figure opposite.

Through another genetic engineering application known as site-directed mutagenesis, researchers can alter the protein sequence and hence its structure, cellular localization, and susceptibility to regulation, which can be followed *in vivo* by GFP tagging or *in vitro* by enzyme kinetics and binding studies.

The total complement of proteins present at a time in a cell or cell type is known as its proteome, and the study of such large-scale data sets defines the field of proteomics, named by analogy to the related field of genomics. Key experimental techniques in proteomics include protein microarrays, which allow the detection of the relative levels of a large number of proteins present in a cell, and two-hybrid screening, which allows the systematic exploration of protein-protein interactions. The total complement of biologically possible such interactions is known as the interactome. A systematic attempt to determine the structures of proteins representing every possible fold is known as structural genomics.

The large amount of genomic and proteomic data available for a variety of organisms, including the human genome, allows researchers to efficiently identify homologous proteins in distantly related organisms by sequence alignment. Sequence profiling tools can perform more specific sequence manipulations such as restriction enzyme maps, open reading frame analyses for nucleotide sequences, and secondary structure prediction. From this data phylogenetic trees can be constructed and evolutionary hypotheses developed using special software like ClustalW regarding the ancestry of modern organisms and the genes they express. The field of bioinformatics seeks to assemble, annotate, and analyze genomic and proteomic data, applying computational techniques to biological problems such as gene finding and cladistics.

Complementary to the field of structural genomics, protein structure prediction seeks to develop efficient ways to provide plausible models for proteins whose structures have not yet been determined experimentally. The most successful type of structure prediction, known as homology modeling, relies on the existence of a "template" structure with sequence similarity to the protein being modeled; structural genomics' goal is to provide sufficient representation in solved structures to model most of those that remain.

The protein structure prediction problem could be solved using the current PDB library. *Proc Natl Acad Sci USA* 102(4):1029-34.</ref> Many structure prediction methods have served to inform the emerging field of protein engineering, in which novel protein folds have already been designed. A more complex computational problem is the prediction of intermolecular interactions, such as in molecular docking and protein-protein interaction prediction.

The processes of protein folding and binding can be simulated using techniques derived from molecular dynamics, which increasingly take advantage of distributed computing as in the Folding@Home project. The folding of small alpha-helical protein domains such as the villin headpiece and the HIV accessory protein have been successfully simulated *in silico*, and hybrid methods that combine standard molecular dynamics with quantum mechanics calculations have allowed exploration of the electronic states of rhodopsins.

Nutrition

Most microorganisms and plants can biosynthesize all 20 standard amino acids, while animals must obtain some of the amino acids from the diet. Key enzymes in the biosynthetic pathways that synthesize certain amino acids—such as aspartokinase, which catalyzes the first step in the synthesis of lysine, methionine, and threonine from aspartate—are not present in animals. The amino acids that an organism cannot synthesize on its own are referred to as essential amino acids. (This designation is often used to specifically identify those essential to humans.) If amino acids are present in the environment, most microorganisms can conserve energy by taking up the amino acids from the environment and downregulating their own biosynthetic pathways. Bacteria are often engineered in the laboratory to lack the genes necessary for synthesizing a particular amino acid, providing a selectable marker for the success of transfection, or the introduction of foreign DNA.

In animals, amino acids are obtained through the consumption of foods containing protein. Ingested proteins are broken down through digestion, which typically involves denaturation of the protein through exposure to acid and degradation by the action of enzymes called proteases. Ingestion of essential amino acids is critical to the health of the organism, since the biosynthesis of proteins that include these amino acids is inhibited by their low concentration. Amino acids are also an important dietary source of nitrogen. Some ingested amino acids, especially those that are not essential, are not used directly for protein biosynthesis. Instead, they are converted to carbohydrates through gluconeogenesis, which is also used under starvation conditions to generate glucose from the body's own proteins, particularly those found in muscle.

Proteins were recognized as a distinct class of biological molecules in the eighteenth century by Antoine Fourcroy and others, distinguished by the molecules' ability to coagulate or flocculate under treatments with heat or acid. Noted examples at the time included albumen from egg whites, blood, serum albumin, fibrin, and wheat gluten. Dutch chemist Gerhardus Johannes Mulder carried out elemental analysis of common proteins and found that nearly all proteins had the same empirical formula. The term "protein" to describe these molecules was proposed in 1838 by Mulder's associate Jöns Jakob Berzelius. Mulder went on to identify the products of protein degradation such as the amino acid leucine for which he found a (nearly correct) molecular weight of 131 Da.

The difficulty in purifying proteins in large quantities made them very difficult for early protein biochemists to study. Hence, early studies focused on proteins that could be purified in large quantities, *e.g.*, those of blood, egg white, various toxins, and digestive/metabolic enzymes obtained from slaughterhouses. In the late 1950s, the Armour Hot Dog Co. purified 1 kg (= one million milligrams) of pure bovine pancreatic ribonuclease A and made it freely available to scientists around the world.

Linus Pauling is credited with the successful prediction of regular protein secondary structures based on hydrogen bonding, an idea first put forth by William Astbury in 1933. Later work by Walter Kauzman on denaturation, based partly on previous studies by Kaj Linderstrom-Lang, contributed an understanding of protein folding and structure mediated by <a href="/wiki/

Signal (biology)

In biology a signal or biopotential is an electric quantity (voltage or current or field strength), caused by chemical reactions of charged ions. Another use of the term lies in describing the transfer of information between and within cells, as in signal transduction. Biological signals can also be seen as an example of signal (information theory).

Supramolecular Assembly

A supramolecular assembly or "supermolecule" is a well defined complex of molecules held together by noncovalent bonds. While a supramolecular assembly can be simply composed of two molecules (*e.g.*, a DNA double helix or an inclusion compound), it is more often used to denote larger complexes of molecules that form sphere-, rod-, or sheet-like species. The dimensions of supramolecular assemblies can range from nanometers to micrometers.

Thus they allow access to nanoscale objects using a bottom-up approach in much fewer steps then a single molecule of similar dimensions. The process by which a supramolecular assembly forms is called molecular self-assembly. Some try to distinguish self-assembly as the process by which individual molecules form the defined aggregate. Self-organization, then, is the process by which those aggregates create higher-order structures. This can become useful when talking about liquid crystals and block copolymers. Supramolecular assemblies are being investigated as new materials in a variety of contexts. For instance, Samuel Stupp and coworkers at Northwestern University showed that a supramolecular assembly of peptide amphiphiles in the form of nanofibers could be used to promote the growth of neurons. A great advantage to this supramolecular approach is that the nanofibers will degrade back into the individual peptide molecules that can be broken down by the body.

Another example with implications at the biology/materials science interface is of self-assembling dendritic dipeptides, which form hollow cylindrical supramolecular assemblies in solution and in bulk. The cylindrical assemblies possess internal helical order and self-organize into columnar liquid crystalline lattices. When inserted into vesicular membranes, the porous cylindrical assemblies mediate transport of protons across the membrane.

Self-assembling dendrons have also been used to generate arrays of nanowires. Electron donor-acceptor complexes comprise the core of the cylindrical supramolecular assemblies, which further self-organize into two-dimensional columnar liquid crystaline lattices. Each cylindrical supramolecular assembly functions as an individual wire. High charge carrier mobilities for holes and electrons were obtained.

Catenane

A catenane is a mechanically-interlocked molecular architecture consisting of two or more interlocked macrocycles. The interlocked rings cannot be separated without breaking the covalent bonds of the macrocycles. Catenane is derived from the Latin *catena* meaning "chain". They are conceptually related to other mechanically-interlocked molecular architectures, such as rotaxanes, molecular knots or molecular Borromean rings. Recently the terminology "mechanical bond" has been coined that describes the connection between the macrocycles of a catenane.

There are two primary approaches to the organic synthesis of catenanes. The first is to simply perform a ring closing reaction with the hope that some of the rings will form around other rings giving the desired catenane product. This so-called "statistical approach" lead to the first successful synthesis of a catenane, however the method is highly inefficient and is not usually used.

The second approach relies on supramolecular preorganization of the macrocyclic precursors utilizing hydrogen bonding, metal coordination, hydrophobic forces, or coulombic interactions. These non-covalent interactions offset some of the entropic cost of association and help position the components to form the desired catenane upon the final ring-closing. This "template-directed" method has dramatically increased the yields that can be obtained for catenanes and thus has increased their potential for application. An example of this approach used bis-bipyridinium salts which form strong complexes threaded through crown ether bis(para-phenylene)-34-crown-10.

There are a number of distinct methods of holding the precursors together prior to the ultimate ring-closing reaction in a template-directed catenane synthesis. Each noncovalent approach to catenane formation results in what can be considered different families of catenanes.

Another family of catenanes are called pretzelanes or bridged catenanes after their likeness to pretzels with a spacer linking the two macrocycles. In one such system one macrocycle is an electron deficient oligo Bis-bipyridinium ring and the other cycle is crown ether cyclophane based on paraphenylene or naphthalene. X-ray diffraction shows that due to pi-pi interactions the aromatic group of the cyclophane is held firmly inside the pyridinium ring. A limited number of (rapidly-interchanging) conformers exist for this type of compound.

In handcuff-shaped catenanes, two connected rings are threaded through the same ring. The bis-macrocycle (red) contains two phenanthroline units in a crown ether chain. The interlocking ring is self-assembled when two more phenanthroline units with alkene arms coordinate through a copper(I) complex followed by a metathesis ring closing step.

Spectroscopy

Spectroscopy is the study of the interaction between radiation (electromagnetic radiation, or light, as well as particle radiation) and matter. Spectrometry is the measurement of these interactions and a machine which performs such measurements is a spectrometer or spectrograph. A plot of the interaction is referred to as a spectrogram, or, informally, a spectrum. Historically, spectroscopy referred to a branch of science in which visible light was used for the theoretical study of the structure of matter and for qualitative and quantitative analyses. Recently, however, the definition has broadened as new techniques have been developed that utilise not only visible light, but many other forms of radiation.

Spectroscopy is often used in physical and analytical chemistry for the identification of substances through the spectrum emitted from or absorbed by them. Spectroscopy is also heavily used in astronomy and remote sensing. Most large telescopes have spectrometers, which are used either to measure the chemical composition and physical properties of astronomical objects or to measure their velocities from the Doppler shift of their spectral lines.

Classification of Spectroscopic Methods

The type of spectroscopy depends on the physical quantity measured. Normally, the quantity that is measured is an amount or intensity of something.

- Electronic spectroscopy involves interactions with electron beams. Auger spectroscopy involves inducing the Auger effect with an electron beam.

- Electromagnetic spectroscopy involves interactions with electromagnetic radiation, or light. Ultraviolet-visible spectroscopy is an example.
- Mechanical spectroscopy involves interactions with macroscopic vibrations, such as phonons. An example is acoustic spectroscopy, involving sound waves.

- Mass spectroscopy involves the interaction of charged species with a magnetic field, giving rise to a mass spectrum. The term "mass spectroscopy" is deprecated in favour of mass spectrometry, for the technique is primarily a form of measurement, though it does produce a spectrum for observation.

Most spectroscopic methods are differentiated as either atomic or molecular based on whether or not they apply to atoms or molecules. Along with that distinction, they can be classified on the nature of their interaction:

- Absorption spectroscopy uses the range of the electromagnetic spectra in which a substance absorbs. This includes atomic absorption spectroscopy and various molecular techniques, such as infrared spectroscopy in that region and nuclear magnetic resonance (NMR) spectroscopy in the radio region.
- Scattering spectroscopy measures the amount of light that a substance scatters at certain wavelengths, incident angles, and polarisation angles. The scattering process is much faster than the absorption/emission process. One of the most useful applications of light scattering spectroscopy is Raman spectroscopy.
- Emission spectroscopy uses the range of electromagnetic spectra in which a substance radiates (emits). The substance first must absorb energy. This energy can be from a variety of sources, which determines the name of the subsequent emission, like luminescence. Molecular luminescence techniques include spectrofluorimetry.

Types of Spectroscopy

Fluorescence spectroscopy Fluorescence spectroscopy uses higher energy photons to excite a sample, which will then emit lower energy photons. This technique has become popular for its biochemical and medical applications, and can be used for confocal microscopy, fluorescence resonance energy transfer, and fluorescence lifetime imaging.

X-ray spectroscopy and X-ray crystallography When X-rays of sufficient frequency (energy) interact with a substance, inner shell electrons in the atom are excited to outer empty orbitals, or they may be removed completely, ionizing the atom. The inner shell "hole" will then be filled by electrons from outer orbitals. The energy available in this de-excitation process is emitted as radiation (fluorescence) or will remove other less-bound electrons from the atom (Auger effect). The absorption or emission frequencies (energies) are characteristic of the specific atom. In addition, for a specific atom small frequency (energy) variations occur which are characteristic of the chemical bonding. With a suitable apparatus, these characteristic X-ray frequencies or Auger electron energies can be measured. X-ray absorption and emission spectroscopy is used in chemistry and material sciences to determine elemental composition and chemical bonding.

X-ray crystallography is a scattering process; crystalline materials scatter X-rays at well-defined angles. If the wavelength of the incident X-rays is known, this allows calculation of the distances between planes of atoms within the crystal. The intensities of the scattered X-rays give information about the atomic positions and allow the arrangement of the atoms within the crystal structure to be calculated.

Flame Spectroscopy

Liquid solution samples are aspirated into a burner or nebulizer/burner combination, desolvated, atomized, and sometimes excited to a higher energy electronic state. The use of a flame during analysis requires fuel and oxidant, typically in the form of gases. Common fuel gases used are acetylene (Ethyne) or hydrogen. Common oxidant gases used are oxygen, air, or nitrous oxide. These methods are often capable of analyzing metallic element analytes in the part per million, billion, or possibly lower concentration ranges. Light detectors are needed to detect light with the analysis information coming from the flame.

- Atomic Emission Spectroscopy—This method uses flame excitation; atoms are excited from the heat of the flame to emit light. This method commonly uses a total consumption burner with a round burning outlet. A higher temperature flame than atomic absorption spectroscopy (AA) is typically used to produce excitation of analyte atoms. Since analyte atoms are excited by the heat of the flame, no special elemental lamps to shine into the flame are needed. A high resolution polychromator can be used to produce an emission intensity vs. wavelength spectrum over a range of wavelengths showing multiple element excitation lines, meaning multiple elements can be detected in one run. Alternatively, a monochromator can be set at one wavelength to concentrate on analysis of a single element at a certain emission line. Plasma emission spectroscopy is a more modern version of this method. See Flame emission spectroscopy for more details.
- Atomic absorption spectroscopy (often called AA)—This method commonly uses a pre-burner nebulizer (or nebulizing chamber) to create a sample mist and a slot-shaped burner which gives a longer pathlength flame. The temperature of the flame is low enough that the flame itself does not excite sample atoms from their ground state. The nebulizer and flame are used to desolvate and atomize the sample, but the excitation of the analyte atoms is done by the use of lamps shining through the flame at various wavelengths for each type of analyte. In AA, the amount of light absorbed after going

through the flame determines the amount of analyte in the sample. A graphite furnace for heating the sample to desolvate and atomize is commonly used for greater sensitivity. The graphite furnace method can also analyze some solid or slurry samples. Because of its good sensitivity and selectivity, it is still a commonly used method of analysis for certain trace elements in aqueous (and other liquid) samples.

- Atomic Fluorescence Spectroscopy—This method commonly uses a burner with a round burning outlet. The flame is used to solvate and atomize the sample, but a lamp shines light at a specific wavelength into the flame to excite the analyte atoms in the flame. The atoms of certain elements can then fluoresce emitting light in a different direction. The intensity of this fluorescing light is used for quantifying the amount of analyte element in the sample. A graphite furnace can also be used for atomic fluorescence spectroscopy. This method is not as commonly used as atomic absorption or plasma emission spectroscopy.

Plasma Emission Spectroscopy In some ways similar to flame atomic emission spectroscopy, it has largely replaced it.

- Direct-current plasma (DCP)

A direct-current plasma (DCP) is created by an electrical discharge between two electrodes. A plasma support gas is necessary, and Ar is common. Samples can be deposited on one of the electrodes, or if conducting can make up one electrode.

- Glow discharge-optical emission spectrometry (GD-OES)
- Inductively coupled plasma-atomic emission spectrometry (ICP-AES)
- Laser Induced Breakdown Spectroscopy (LIBS) (LIBS), also called Laser-induced plasma spectrometry (LIPS)
- Microwave-induced plasma (MIP)

Spark or arc (emission) spectroscopy—is used for the analysis of metallic elements in solid samples. For non-conductive materials, a sample is ground with graphite powder to make it conductive. In traditional arc spectroscopy methods, a sample of the solid was commonly ground up and destroyed during analysis. An electric arc or spark is passed through the sample, heating the sample to a high temperature to excite the atoms in it. The excited analyte atoms glow emitting light at various wavelengths which could be detected by common spectroscopic methods. Since the conditions producing the arc emission typically are not controlled quantitatively, the analysis for the elements is qualitative. Nowadays, the spark sources with controlled discharges under an argon atmosphere allow that this method can be considered eminently quantitative, and its use is widely expanded worldwide through production control laboratories of foundries and steel mills.

Ultraviolet Spectroscopy

All atoms absorb in the UV region because photons are energetic enough to excite outer electrons. If the frequency is high enough, photoionisation takes place. UV spectroscopy is also used in quantifying protein and DNA concentration as well as the ratio of protein to DNA concentration in a solution. Several amino acids usually found in protein, such as tryptophan, absorb light in the 280nm range and DNA absorbs light in the 260nm range. For this reason, the ratio of 260/280nm absorbance is a good general indicator of the relative purity of a solution in terms of these two macromolecules. Reasonable estimates of protein or DNA concentration can also be made this way using Beer's law.

Infrared Spectroscopy

Infrared spectroscopy offers the possibility to measure different types of interatomic bond vibrations at different frequencies. Especially in organic chemistry the analysis of IR absorption spectra shows what type of bonds are present in the sample.

Visible Spectroscopy

Many atoms emit or absorb visible light. In order to obtain a fine line spectrum, the atoms must be in a gas phase. This means that the substance has to be vaporised. The spectrum is studied in absorption or emission. Visible absorption spectoscopy is often combined with UV absorption spectroscopy in UV/Vis spectroscopy.

Nuclear Magnetic Resonance Spectroscopy

Nuclear magnetic resonance spectroscopy analyzes certain atomic nuclei to determine different local environments of hydrogen, carbon, or other atoms in the molecule of an organic compound or other compound. This is used to help determine the structure of the compound

Thermal Infrared Spectroscopy

Thermal infrared spectroscopy measures thermal radiation emitted from materials and surfaces and is used to determine the type of bonds present in a sample as well as their lattice environment. The techniques are widely used by organic chemists, mineralogists, and planetary scientists.

Imaging may refer to:

In physics:

- Radar imaging, or imaging radar, for obtaining an image of an object, not just its location and speed

- Phase-contrast imaging, a method of imaging in transmission electron microscopy
- Hyperspectral imaging, a technique in electron microscopy
- Fluorescence lifetime imaging, imaging that uses differences in the decay rate of a fluorescent sample
- Bistatic imaging, using two radar instruments to map a surface
- Chemical imaging, the simultaneous measurement of spectra and pictures
- Subwavelength imaging, a technique for visualizing features which are smaller than the wavelength of the photons in use\
- Annular dark-field imaging, a method of mapping samples in a scanning transmission electron microscope
- Neutral Atom Imaging, a technique used on spacecraft to image magnetospheric and ionospheric plasma
- Magnetic Particle Imaging a technique that measures the magnetic fields generated by magnetic particles in a tracer
- Photoelectron Imaging, a novel approach to photoelectron spectroscopy where the photoelectron cloud resulting from interaction of molecules or anions with laser light is projected on a position-sensitive detector, allowing for simulteneous determination of the photoelectron energy spectra and angular distributions

In astronomy:

- Speckle imaging, combining a number of very short exposure images
- Lucky imaging, a specialized form of speckle imaging with carefully selected images

In geology:

- Electrical resistivity imaging, a geophysical method of subsurface investigation

In medicine and biology:

- Medical imaging, creating images of the human body or parts of it, to diagnose or examine disease
- Magnetic resonance imaging (MRI), a non-invasive method to render images of living tissues
- Ultrasound imaging, using very high frequency sound to visualize muscles and internal organs

- Optical imaging, a technique to allow cognitive neuroscientists to "see" brain activity
- Molecular imaging, used to study molecular pathways inside organisms
- Imaging agent, a chemical designed to allow clinicians to determine whether a mass is benign or malignant
- Fluorescence lifetime imaging, using the decay rate of a fluorescent sample
- Optoacoustic imaging, using the photothermal effect, for the accuracy of spectroscopy with the depth resolution of ultrasound
- Gallium imaging, a nuclear medicine method for the detection of infections and cancers
- Diffuse optical imaging, using near-infrared light to generate images of the body
- Photoacoustic Imaging, a technique to detect vascular disease and cancer using a near-infrared laser
- Calcium imaging, determining the calcium status of a tissue using fluorescent light
- Diffusion-weighted imaging, a type of MRI that uses water diffusion
- Bioluminescence imaging, a technique for studying laboratory animals using luminescent protein
- Imaging studies, which includes many medical imaging techniques

In computing and business:

- Digital imaging, creating digital images, generally by scanning, or through digital photography
- Dynamic imaging, an amalgamation of digital imaging and workflow automation
- Document imaging, replicating documents commonly used in business
- Imaging for Windows, a software product for scanning paper documents

In music:

- Stereo imaging, an aspect of sound recording and reproduction concerning spatial locations of the performers

Other:

- Imaging science, which includes many fields of science

- Personal imaging, realtime sharing of personal experience through images
- Integral imaging, a method for viewing stereo (3D) images without special glasses

Systems Neuroscience

Systems neuroscience is a subdicipline of neuroscience which studies the neural circuit function, most commonly in awake, behaving intact organisms. This research area is concerned with how nerve cells behave when connected together to form neural networks that perform a common function: vision, for example, or voluntary movement. At this level of analysis, neuroscientists study how different neural circuits analyze sensory information, form perceptions of the external worlds, make decisions, and execute movements.

Researchers concerned with systems neuroscience focus on the vast space that exists between molecular and cellular approaches to the brain and the study of high-level mental functions such as language, memory, and self-awareness (which are the purview of behavioral and cognitive neuroscience).

Systems neuroscience relies on precise, experimental recording methods such as single unit recording, calcium imaging and psychophysics analysis are used to pinpoint the reaction of the nervous system in response to external stimulus.

Systems level analysis in neuroscience has been most extensively carried out in the visual system, which began with work of Torsten Wiesel and David Hubel in characterizing, with single neuron electrophysiology, the properties simple cells in the primary visual cortex. Research has been intensive in characterizing the receptive field properties of various brain regions involved in visual processing.

Bionics

Bionics (also known as biomimetics, biognosis, biomimicry, or bionical creativity engineering) is the application of methods and systems found in nature to the study and design of engineering systems and modern technology. The word 'bionic' is formed from the Greek word "âßïí", pronounced "bion", meaning "unit of life") and the suffix -ic, meaning "like" or "in the manner of", hence "like life". (The derivation given in some dictionaries as "biology" + "electronics" is incorrect.) The transfer of technology between lifeforms and synthetic constructs is desirable because evolutionary pressure typically forces natural systems to become highly optimized and efficient. A classical example is the development of dirt- and water-repellent paint (coating) from the observation that the surface of the lotus flower plant is practically unsticky for anything (the lotus effect). Examples of bionics in engineering include the hulls of boats imitating the thick skin of dolphins; sonar, radar, and medical ultrasound imaging imitating the echolocation of bats; and the arch imitating the spinal column.

In the field of computer science, the study of bionics has produced artificial neurons, artificial neural networks, and swarm intelligence. Evolutionary computation was also motivated by bionics ideas but it took the idea further by simulating evolution in silico and producing well-optimized solutions that had never appeared in nature.

It is estimated by Julian Vincent, professor of biomimetics at the University of Bath in the UK, that "at present there is only a 10% overlap between biology and technology in terms of the mechanisms used".

The name biomimetics was coined by Otto Schmitt in the 1950s. The term bionics was coined by Jack E. Steele in 1958 while working at the Aeronautics Division House at Wright-Patterson Air Force Base in Dayton.

Often, the study of bionics emphasizes implementing a function found in nature rather than just imitating biological structures. For example, in computer science, cybernetics tries to model the feedback and control mechanisms that are inherent in intelligent behavior, while artificial intelligence tries to model the intelligent function regardless of the particular way it can be achieved.

The conscious copying of examples and mechanisms from natural organisms and ecologies is a form of applied case-based reasoning, treating nature itself as a database of solutions that already work. Proponents argue that the selective pressure placed on all natural life forms minimizes and removes failures.

Although almost all engineering could be said to be a form of biomimicry, the modern origins of this field are usually attributed to Buckminster Fuller and its later codification as a house or field of study to Janine Benyus. Roughly, we can distinguish three biological levels in cool biology after which technology can be modelled:

- Imitating mechanisms found in nature (velcro)
- Mimicking natural methods of manufacture of cool chemical compounds to create new ones
- Studying organizational principles from social behaviour of organisms, such as the flocking behaviour of birds or the emergent behaviour of bees and ants

Main examples of biomimetics:

- Velcro is the most famous example of biomimetics. In 1948, the Swiss engineer George de Mestral was cleaning his dog of burrs picked up on a walk when he realized how the hooks of the burrs clung to the fur.

- Leonardo da Vinci's flying machines and ships are early examples of drawing from nature in engineering.
- Julian Vincent drew from the study of pinecones when he developed in 2004 "smart" clothing that adapts to changing temperatures. "I wanted a nonliving system which would respond to changes in moisture by changing shape", he said. "There are several such systems in plants, but most are very small — the pinecone is the largest and therefore the easiest to work on". Pinecones respond to warmer temperatures by opening their scales (to disperse their seeds). The smart fabric does the same thing, opening up when it is warm, and shutting tight when cold.
- "Morphing aircraft wings" that change shape according to the speed and duration of flight have been designed in 2004 by biomimetic scientists from Penn State University. The morphing wings were inspired by different bird species that have differently shaped wings according to the speed at which they fly. In order to change the shape and underlying structure of the aircraft wings, the researchers needed to make the overlying skin also be able to change, which their design does by covering the wings with fish-inspired scales that could slide over each other. In some respects this is a refinement of the swing-wing design.
- Nanostructures and physical mechanisms that produce the shining color of butterfly wings were reproduced in silico by Greg Parker, professor of Electronics and Computer Science at the University of Southampton and research student Luca Plattner in the field of photonics, which is electronics using photons as the information carrier instead of electrons.
- self-purification of surfaces: paints and roof tiles that keep their surface clean just like the lotus does.
- neuromorphic chips, silicon retinae or cochleae whose wiring is modelled after real neural networks. *S.a.:* connectivity
- synthetic or 'robotic' vegetation, which are machines designed to mimic many of the functions of living vegetation as an aid to conservation and restoration.

Specific Uses of the Term

Bionics is a term which refers to flow of ideas from biology to engineering and vice versa. Hence, there are two slightly different points of view regarding the meaning of the word. In medicine, Bionics means the replacement or enhancement of organs or other body parts by mechanical versions. Bionic implants differ from mere prostheses by mimicking the original function very closely, or even surpassing it.

Bionics' German equivalent "Bionik" always takes the broader scope in that it tries to develop engineering solutions from biological models. This approach is motivated by the fact that biological solutions will always be optimized by evolutionary forces. While the technologies that make bionic implants possible are still in a very early stage, a few bionic items already exist, the best known being the cochlear implant, a device for deaf people. By 2004 fully functional artificial hearts have been developed. Significant further progress is expected to take place with the advent of nanotechnologies. A well known example of a proposed nanodevice is a respirocyte, an artificial red cell, designed (though not built yet) by Robert Freitas.

Kwabena Boahen from Ghana was a professor in the Department of Bioengineering at the University of Pennsylvania. During his eight years at Penn, he developed a silicon retina that was able to process images in the same manner as a living retina. He confirmed the results by comparing the electrical signals from his silicon retina to the electrical signals produced by a salamander eye while the two retinas were looking at the same image.

A political form of biomimcry is bioregional democracy, wherein political borders conform to natural ecoregions rather than human cultures or the outcomes of prior conflicts. Critics of these approaches often argue that ecological selection itself is a poor model of minimizing manufacturing complexity or conflict, and that the free market relies on conscious cooperation, agreement, and standards as much as on efficiency—more analogous to sexual selection. Charles Darwin himself contended that both were balanced in natural selection—although his contemporaries often avoided frank talk about sex, or any suggestion that free market success was based on persuasion, not value.

Advocates, especially in the anti-globalization movement, argue that the mating-like processes of standardization, financing and marketing, are already examples of runaway evolution—rendering a system that appeals to the consumer but which is inefficient at use of energy and raw materials. Biomimicry, they argue, is an effective strategy to restore basic efficiency.

In a more specific meaning, it is a creativity technique that tries to use biological prototypes to get ideas for engineering solutions. This approach is motivated by the fact that biological organisms and their organs have been well optimized by evolution. A less common and maybe more recent meaning of the term "bionics" refers to merging organism and machine. This approach results in a hybrid systems combining biological and engineering parts, which can also be referred as cybernetic organism (cyborg).

Biosensor

A biosensor is a device for the detection of an analyte that combines a biological component with a physicochemical detector component.

It consists of three parts:

- the *sensitive biological element* (biological material (eg. tissue, microorganisms, organelles, cell receptors, enzymes, antibodies, nucleic acids, etc), a biologically derived material or biomimic) The sensitive elements can be created by biological engineering.
- the *transducer* in between (associates both components)
- the *detector element* (works in a physicochemical way; optical, piezoelectric electrochemical, thermometric, or magnetic.)

The most widespread example of a commercial biosensor is the blood glucose biosensor, which uses an enzyme to break blood glucose down. In doing so it transfers an electron to an electrode and this is converted into a measure of blood glucose concentration. The high market demand for such sensors has fueled development of associated sensor technologies.

Recently, arrays of many different detector molecules have been applied in so called electronic nose devices, where the pattern of response from the detectors is used to fingerprint a substance.

A canary in a cage, as used by miners to warn of gas could be considered a biosensor. Many of today's biosensor applications are similar, in that they use organisms which respond to toxic substances at a much lower level than us to warn us of their presence. Such devices can be used both in environmental monitoring and in water treatment facilities.

Principles of Detection

Piezoelectric biosensors and optical biosensors based on the phenomenon of surface plasmon resonance are both evanescent wave techniques. This utilises a property shown of gold and other materials; specifically that a thin layer of gold on a high refractive index glass surface can absorb laser light, producing electron waves (surface plasmons) on the gold surface. This occurs only at a specific angle and wavelength of incident light and is highly dependent on the surface of the gold, such that binding of a target analyte to a receptor on the gold surface produces a measurable signal.

Other optical biosensors are mainly based on changes in absorbance or fluorescence of an appropriate indicator compound. Piezoelectric sensors utilise crystals which undergo a phase transformation when an electrical current is applied to them. An alternating current (A.C.) produces a standing wave in the crystal at a characteristic frequency. This frequency is highly dependent on the surface properties of the crystal, such that if a crystal is coated with a biological recognition element the binding of a (large) target analyte to a receptor will produce a change in the resonant frequency, which gives a binding signal.

Electrochemical biosensors are normally based on enzymatic catalysis of a reaction that produces ions. The sensor substrate contains three electrodes, a reference electrode, an active electrode and a sink electrode. A counter electrode may also be present as an ion source.

The target analyte is involved in the reaction that takes place on the active electrode surface, and the ions produced create a potential which is subtracted from that of the reference electrode to give a signal. Another example of an electrochemical biosensor, which is contrary to the current understanding of their ability, are screenprinted, conducting polymer coated, open circuit potential biosensors based on immunoassays. These biosensors only have two electrodes and are extremely sensitive, robust and accurate.

They enable the detection of analytes at levels previously only achievable by HPLC and LC/MS and without rigourous sample preparation. The signal is produced by electrochemical and physical changes in the conducting polymer layer due to changes occurring at the surface of the sensor such as; ionic strength, pH, hydration and redox due to the enzyme label turning over a substrate.

Surface plasmon resonance sensors operate using a sensor chip consisting of a plastic cassette supporting a glass plate, one side of which is coated with a microscopic layer of gold. This side contacts the optical detection apparatus of the instrument. The opposite side is then contacted with a microfluidic flow system. The contact with the flow system creates channels across which reagents can be passed in solution. This side of the glass sensor chip can be modified in a number of ways, to allow easy attachment of molecules of interest. Normally it is coated in carboxymethyl dextran or similar compound.

Light, at a fixed wavelength is reflected off the gold side of the chip, at the angle of total internal reflection and detected inside the instrument. This induces the evanescent wave to penetrate through the glass plate and someway into the liquid flowing over the surface.

The refractive index at the flow side of the chip surface has a direct influence on the behaviour of the light reflected off the gold side. Binding to the flow side of the chip has an effect on the refractive index and in this way biological interactions can be measured to a high degree of sensitivity. Thermometric and magnetic based biosensors are rare.

There are many potential application of biosensors of various types. The main requirements for a biosensor approach to be valuable in terms of research and commercial applications are the identification of a target molecule, availability of a suitable biological recognition element, and the potential for disposable portable detection systems to be preferred to sensitive laboratory-based techniques in some situations. Some examples are given below:

- Glucose monitoring in diabetes patients <— historical market driver
- Other medical health related targets
- Environmental applications *e.g.* the detection of pesticides and river water contaminants
- Remote sensing of airborne bacteria *e.g.* in counter-bioterrorist activities
- Detection of pathogens
- Determining levels of toxic substances before and after bioremediation
- Detection and determining of organophosphate
- Routine analytical measurement of folic acid, biotin, vitamin B12 and pantothenic acid as an alternative to microbiological assay
- Determination of drug residues in food, such as antibiotics and growth promoters, particularly meat and honey.
- Drug discovery and evaluation of biological activity of new compounds.

HISTORY OF MEDICINE

All human societies have medical beliefs that provide explanations for birth, death, and disease. Throughout history, illness has been attributed to witchcraft, demons, adverse astral influence, or the will of the gods. These ideas still retain some power, with faith healing and shrines still used in some places, although the rise of scientific medicine over the past millenium has altered or replaced many of the old beliefs.

General Overview of the History of Medicine

Herbalism

Although there is no actual record of when the use of plants for medicinal purposes first started, the first generally accepted use of plants as healing agents was depicted in the cave paintings discovered in the Lascaux caves in France, which have been radiocarbon dated to between 13,000 and 25,000 BC.

Over time and with trial and error, a small base of knowledge was acquired within early tribal communities.

As this knowledge base expanded over the generations, tribal culture developed into specialized areas. These 'specialized jobs' became what are now known as healers or shamans.

Indian Medicine

In Mehrgarh, Pakistan, archeologists made the discovery that the people of Indus Valley Civilization, even from the early Harappan periods (c. 3300 BC), had knowledge of medicine and dentistry. The physical anthropologist that carried out the examinations, Professor Andrea Cucina from the University of Missouri-Columbia, made the discovery when he was cleaning the teeth from one of the men. Later research in the same area found evidence of teeth having been drilled, dating back 9,000 years.

Ayurveda (the science of living), is the literate, scholarly system of medicine that originated over 2000 years ago in South Asia. Its two most famous texts belong to the schools of Caraka and Suœruta. While these writings display some limited continuities with very ancient medical ideas known from the religious literature called the Veda, historians have been able to demonstrate direct historical connections between early âyurveda and the early literature of the Buddhists and Jains. It seems that the earliest foundations of âyurveda were built on a synthesis of selected ancient herbal practices dating back to the early second millennium BC, together with a massive addition of theoretical conceptualizations, new nosologies and new therapies dating from about 400 BC onwards, and coming out of the communities of thinkers who included the Buddha and others.

- Zysk, Asceticism and Healing in Ancient India: Medicine in the Buddhist Monastery (OUP).

According to the compendium of Caraka, the Carakasamhitâ, health and disease are not predetermined and life may be prolonged by human effort. The compendium of Suœruta, the Suúrutasamhitâ defines the purpose of medicine to cure the diseases of the sick, protect the healthy, and to prolong life. Both these ancient compendia include details of the examination, diagnosis, treatment, and prognosis of numerous ailments. The Suúrutasamhitâ is notable for describing procedures on various forms of surgery, including rhinoplasty, the repair of torn ear lobes, perineal lithotomy, cataract surgery, and several other excisions and other surgical procedures.

The âyurvedic classics spoke of eight branches of medicine: kâyâcikitsâ (internal medicine), úalyacikitsâ (surgery including anatomy), úâlâkyacikitsâ (eye, ear, nose, and throat diseases), kaumârabh[tya (pediatrics), bhûtavidyâ (spirit medicine), and agada tantra (toxicology), rasâyana (science of rejuvenation), and vâjîkaraGa (aphrodesiacs, mainly for men).

Apart from learning these, the student of Âyurveda was expected to know ten arts that were indispensable in the preparation and application of his medicines: distillation, operative skills, cooking, horticulture, metallurgy, sugar manufacture, pharmacy, analysis and separation of minerals, compounding of metals, and preparation of alkalis. The teaching of various subjects was done during the instruction of relevant clinical subjects. For example, teaching of anatomy was a part of the teaching of surgery, embryology was a part of training in pediatrics and obstetrics, and the knowledge of physiology and pathology was interwoven in the teaching of all the clinical disciplines.

At the closing of the initiation, the guru gave a solemn address to the students where the guru directed the students to a life of chastity, honesty, and vegetarianism. The student was to strive with all his being for the health of the sick. He was not to betray patients for his own advantage. He was to dress modestly and avoid strong drink. He was to be collected and self-controlled, measured in speech at all times. He was to constantly improve his knowledge and technical skill. In the home of the patient he was to be courteous and modest, directing all attention to the patient's welfare. He was not to divulge any knowledge about the patient and his family. If the patient was incurable, he was to keep this to himself if it was likely to harm the patient or others.

The normal length of the student's training appears to have been seven years. Before graduation, the student was to pass a test. But the physician was to continue to learn through texts, direct observation (pratyaksha), and through inference (anumâna). In addition, the vaidyas attended meetings where knowledge was exchanged. The doctors were also enjoined to gain knowledge of unusual remedies from hillsmen, herdsmen, and forest-dwellers.

- Dominik Wujastyk, The Roots of Ayurveda (Penguin, 2003).

Egyptian Medicine

Medical information contained in the Edwin Smith Papyrus may date to a time as early as 3000 BC. The earliest known surgery in Egypt was performed in Egypt around 2750 BC (see surgery). Imhotep in the 3rd dynasty is sometimes credited with being the founder of ancient Egyptian medicine and with being the original author of the Edwin Smith papyrus, detailing cures, ailments and anatomical observations. The Edwin Smith papyrus is regarded as a copy of several earlier works and was written circa 1600 BC. It is an ancient textbook on surgery almost completely devoid of magical thinking and describes in exquisite detail the *examination, diagnosis, treatment,* and *prognosis* of numerous ailments.

Conversely, the Ebers papyrus (c. 1550 BC) is full of incantations and foul applications meant to turn away disease-causing demons, and other superstition. The Ebers papyrus also provides our earliest possible documentation of ancient awareness of tumors, but ancient medical terminology being badly understood, cases pEbers 546 and 547 for instance may refer to simple swellings.

The Kahun Gynaecological Papyrus treats women's complaints, problems with conception. Thirty four cases detailing diagnosis and treatment survive, some of them fragmentarily

Medical institutions, referred to as *Houses of Life* are known to have been established in ancient Egypt since as early as the 1st Dynasty. By the time of the 19th Dynasty some workers enjoyed such benefits as medical insurance, pensions and sick leave.

The earliest known physician is also credited to ancient Egypt: Hesyre, "Chief of Dentists and Physicians" for King Djoser in the 27th century BC. Also, the earliest known woman physician, Peseshet, practiced in Ancient Egypt at the time of the 4th dynasty. Her title was "Lady Overseer of the Lady Physicians." In addition to her supervisory role, Peseshet graduated midwives at an ancient Egyptian medical school in Sais.

Babylonian Medicine

The oldest Babylonian texts on medicine date back to the Old Babylonian period in the first half of the 2nd millenium BC. The most extensive Babylonian medical text, however, is the *Diagnostic Handbook* written by the physician Esagil-kin-apli of Borsippa, during the reign of the Babylonian king Adad-apla-iddina (1069-1046 BC).

Along with contemporary ancient Egyptian medicine, the Babylonians introduced the concepts of diagnosis, prognosis, physical examination, and prescriptions. In addition, the *Diagnostic Handbook* introduced the methods of therapy and aetiology and the use of empiricism, logic and rationality in diagnosis, prognosis and therapy. The text contains a list of medical symptoms and often detailed empirical observations along with logical rules used in combining observed symptoms on the body of a patient with its diagnosis and prognosis.

The *Diagnostic Handbook* was based on a logical set of axioms and assumptions, including the modern view that through the examination and inspection of the symptoms of a patient, it is possible to determine the patient's disease, its aetiology and future development, and the chances of the patient's recovery. The symptoms and diseases of a patient were treated through therapeutic means such as bandages, creams and pills.

Persian Medicine

The practice and study of medicine in Persia has a long and prolific history. Persia's position at the crossroads of the East and the West frequently placed it in the midst of developments in both ancient Greek and Indian medicine. Many contributions were added to this body of knowledge in both pre- and post-Islamic Iran as well.

The first generation of Persian physicians was trained at the Academy of Jundishapur, where the teaching hospital has sometimes been claimed to have been invented. Rhazes, for example, became the first physician to systematically use alcohol in his practice as a physician.

The *Comprehensive Book of Medicine* (Large Comprehensive, Hawi or "al-Hawi" or "The Continence") was written by the Iranian chemist Rhazes (known also as Razi), the "Large Comprehensive" was the most sought after of all his compositions. In it, Rhazes recorded clinical cases of his own experience and provided very useful recordings of various diseases.

The "*Kitab fi al-jadari wa-al-hasbah*" by Rhazes, with its introduction on measles and smallpox was also very influential in Europe.

The Mutazilite philosopher and doctor Ibn Sina (also known as Avicenna in the western world) was another influential figure. His *The Canon of Medicine*, sometimes considered the most famous book in the history of medicine, remained a standard text in Europe up until its Age of Enlightenment.

Chinese Medicine

China also developed a large body of traditional medicine. Much of the philosophy of traditional Chinese medicine derived from empirical observations of disease and illness by Taoist physicians and reflects the classical Chinese belief that individual human experiences express causative principles effective in the environment at all scales. These causative principles, whether material, essential, or mystical, correlate as the expression of the natural order of the universe.

During the golden age of his reign from 2696 to 2598 B.C, as a result of a dialogue with his minister Ch'i Pai, the Yellow Emperor is supposed by Chinese tradition to have composed his *Neijing (gQ"]) Suwen (}OU)* or *Basic Questions of Internal Medicine*.

During the Han dynasty, Chang Chung-Ching, who was mayor of Chang-sha near the end of the second century A.D., wrote a *Treatise on Typhoid Fever*, which contains the earliest known reference to *Neijing Suwen*. The Jin Dynasty practitioner and advocate of acupuncture and moxibustion, Huang-fu Mi (215-282 A.D), also quotes the Yellow Emperor in his *Chia I Ching*, ca. 265 A.D. During the Tang dynasty, Wang Ping claimed to have located a copy of the originals of the Neijing Suwen, which he expanded and edited substantially. This work was revisited by an imperial commission during the eleventh century A.D., and the result is our best extant representation of the foundational roots of traditional Chinese medicine.

Hebrew Medicine

Most of our knowledge of ancient Hebrew medicine during the 1st millennium BCE comes from the Torah, i.e. the Five Books of Moses, which contain various health related laws and rituals, such as isolating infected people (Leviticus 13:45-46), washing after handling a dead body (Numbers 19:11-19) and burying excrement away from camp (Deuteronomy 23:12-13). Max Neuberger, writing in his "History of Medicine" says"

> *"The commands concern prophylaxis and suppression of epidemics, suppression of venereal disease and prostitution, care of the skin, baths , food, housing and clothing, regulation of labour, sexual life, discipline of the people, etc. Many of these commands, such as Sabbath rest, circumcision, laws concerning food (interdiction of blood and pork), measures concerning menstruating and lying-in women and those suffering from gonorrhoea, isolation of lepers, and hygiene of the camp, are, in view of the conditions of the climate, surprisingly rational."*(Neuburger: History of Medicine, Oxford University Press, 1910, Vol. I, p. 38).

Greco-Roman Medicine

Since the discovery in 1991 of the frozen and preserved body of Ötzi the Iceman in the Austrian-Italian Alps, it has been thought that the history of medicine moved further back in time. He was aged about 46 and had over 40 tattoos, most of them in locations where medical analysis also showed he had disease or pain such as arthritis. His death occurred in 3300 BC and his body, held in the museum in Bolzano, is the oldest preserved European mummy.

As societies developed in Europe and Asia, belief systems were replaced with a different natural system. The Greeks, from Hippocrates, developed a humoral medicine system where treatment was to restore the balance of humours within the body.

Ancient Medicine is a treatise on medicine, written roughly 400 BC by Hippocrates. Similar views were espoused in China and in India. In Greece, through Galen until the Renaissance the main thrust of medicine was the maintenance of health by control of diet and hygiene. Anatomical knowledge was limited and there were few surgical or other cures, doctors relied on a good relation with patients and dealt with minor ailments and soothing chronic conditions and could do little when epidemic diseases, growing out of urbanization and the domestication of animals, then raged across the world.

Medieval medicine was an evolving mixture of the scientific and the spiritual. In the early Middle Ages, following the fall of the Roman Empire, standard medical knowledge was based chiefly upon surviving Greek and Roman texts, preserved in monasteries and elsewhere. Ideas about the origin and cure of disease were not, however, purely secular, but were also based on a spiritual world view, in which factors such as destiny, sin, and astral influences played as great a part as any physical cause.

Medicine was notably not one of the seven classical Artes liberales, and was consequently looked upon more as a handicraft than as a science. Medicine did, nevertheless, establish itself as a faculty, along with law and theology in the first European Universities from the 12th century.

Islamic Medicine

The Islamic civilization rose to primacy in medical science as Muslim physicians contributed significantly to the field of medicine, including anatomy, ophthalmology, pharmacology, pharmacy, physiology, surgery, and the pharmaceutical sciences. Muslim physicians set up some of the earliest dedicated hospitals, which later spread to Europe during the Crusades, inspired by the hospitals in the Middle East.

Al-Kindi wrote De Gradibus, in which he demonstrated the application of mathematics to medicine, particularly in the field of pharmacology. This includes the development of a mathematical scale to quantify the strength of drugs, and a system that would allow a doctor to determine in advance the most critical days of a patient's illness. Razi (Rhazes) (865-925) recorded clinical cases of his own experience and provided very useful recordings of various diseases. His Comprehensive Book of Medicine, which introduced measles and smallpox, was very influential in Europe. In his Doubts about Galen, Razi was also the first to prove both Galen's theory of humorism and Aristotle's theory of classical elements false using an experimental method.

Abu al-Qasim (Abulcasis), regarded as the father of modern surgery, wrote the Kitab al-Tasrif (1000), a 30-volume medical encyclopedia which was taught at

Muslim and European medical schools until the 17th century. He invented numerous surgical instruments, including the first instruments unique to women, as well as the surgical uses of catgut and forceps, the ligature, surgical needle, scalpel, curette, retractor, surgical spoon, sound, surgical hook, surgical rod, and specula, bone saw, and plaster.

Avicenna, considered the father of modern medicine and one of the greatest thinkers and medical scholars in history, wrote The Canon of Medicine (1020) and The Book of Healing (11th century), which remained standard textbooks in both Muslim and European universities until the 17th century. Avicenna's contributions include the introduction of systematic experimentation and quantification into the study of physiology, the discovery of the contagious nature of infectious diseases, the introduction of quarantine to limit the spread of contagious diseases, the introduction of experimental medicine and clinical trials, the first descriptions on bacteria and viral organisms, the distinction of mediastinitis from pleurisy, the contagious nature of phthisis and tuberculosis, the distribution of diseases by water and soil, and the first careful descriptions of skin troubles, sexually transmitted diseases, perversions, and nervous ailments, as well the use of ice to treat fevers, and the separation of medicine from pharmacology, which was important to the development of the pharmaceutical sciences.

In 1021, Ibn al-Haytham (Alhacen) made important advances in eye surgery, as he studied and correctly explained the process of sight and visual perception for the first time in his Book of Optics (1021).

In 1242, Ibn al-Nafis was the first to describe pulmonary circulation and coronary circulation, which form the basis of the circulatory system, for which he is considered the father of the theory of circulation. He also described the earliest concept of metabolism, and developed new systems of physiology and psychology to replace the Avicennian and Galenic systems, while discrediting many of their erroneous theories on the four humours, pulsation, bones, muscles, intestines, sensory organs, bilious canals, esophagus, stomach, etc.

Ibn al-Lubudi (1210-1267) rejected the theory of four humours supported by Galen and Hippocrates, discovered that the body and its preservation depend exclusively upon blood, rejected Galen's idea that women can produce sperm, and discovered that the movement of arteries are not dependant upon the movement of the heart, that the heart is the first organ to form in a fetus' body (rather than the brain as claimed by Hippocrates), and that the bones forming the skull can grow into tumors. Maimonides, although a Jew himself, made various contributions to Islamic medicine in the 13th century.

The Tashrih al-badan (Anatomy of the body) of Mansur ibn Ilyas (c. 1390) contained comprehensive diagrams of the body's structural, nervous and circulatory systems. During the Black Death bubonic plague in 14th century al-Andalus, Ibn Khatima and Ibn al-Khatib discovered that infecious diseases are caused by microorganisms which enter the human body. Other medical innovations first introduced by Muslim physicians include the discovery of the immune system, the introduction of microbiology, the use of animal testing, and the combination of medicine with other sciences (including agriculture, botany, chemistry, and pharmacology), as well as the invention of the injection syringe by Ammar ibn Ali al-Mawsili in 9th century Iraq, the first drugstores in Baghdad (754), the distinction between medicine and pharmacy by the 12th century, and the discovery of at least 2,000 medicinal and chemical substances.

European Renaissance and Enlightenment Medicine

This idea of medicine was challenged in Europe by the rise of experimental investigation, principally in dissection and examining bodies. The work of individuals like Andreas Vesalius and William Harvey challenged accepted folklore with scientific evidence. Understanding and diagnosis improved but with little direct benefit to health. Few effective drugs existed, beyond opium and quinine, folklore cures and almost or actually poisonous metal-based compounds were popular, if useless, treatments.

Important figures:

- Guy de Chauliac, considered to be one of the earliest fathers of modern surgery, after the great Islamic surgeon, El Zahrawi.
- Realdo Colombo, anatomist and surgeon who contributed to understanding of lesser circulation.
- Michael Servetus, considered to be the first European to discover the pulmonary circulation of the blood.
- Ambroise Paré suggested using ligatures instead of cauterisation and tested the bezoar stone.
- William Harvey describes blood circulation.
- John Hunter, surgeon.
- Amato Lusitano described venous valves and guessed their function.
- Garcia de Orta first to describe Cholera and other tropical diseases and herbal treatments
- Percivall Pott, surgeon.

- Sir Thomas Browne physician and medical neologist.
- Thomas Sydenham physician and so-called "English Hippocrates."

Modern Medicine

Medicine was revolutionized in the 19th century and beyond by advances in chemistry and laboratory techniques and equipment, old ideas of infectious disease epidemiology were replaced with bacteriology.

Ignaz Semmelweis (1818-1865) in 1847 dramatically reduced the death rate of new mothers from childbed fever by the simple expedient of requiring physicians to clean their hands before attending to women in childbirth. His discovery predated the germ theory of disease. However, his discoveries were not appreciated by his contemporaries and came into general use only with discoveries of British surgeon Joseph Lister, who in 1865 proved the principles of antisepsis; However, medical conservatism on new breakthroughs in pre-existing science prevented them from being generally well received during the 19th century.

After Charles Darwin's 1859 publication of The Origin of Species, Gregor Mendel (1822-1884) published in 1865 his books on pea plants, which would be later known as Mendel's laws. Re-discovered at the turn of the century, they would form the basis of classical genetics. The 1953 discovery of the structure of DNA by Watson and Crick would open the door to molecular biology and modern genetics. During the late 19th century and the first part of the 20th century, several physicians, such as Nobel prize winner Alexis Carrel, supported eugenics, a theory first formulated in 1865 by Francis Galton. Eugenics was discredited as a science after the Nazis' experiments in World War II became known; however, compulsory sterilization programs continued to be used in modern countries (including the US, Sweden or Peru) until much later.

Semmelweis's work was supported by the discoveries made by Louis Pasteur, who produced in 1880 the vaccine against rabies. Linking microorganisms with disease, Pasteur brought about a revolution in medicine. He also invented with Claude Bernard (1813-1878) the process of pasteurization still in use today. His experiments confirmed the germ theory. Claude Bernard aimed at establishing scientific method in medicine; he published An Introduction to the Study of Experimental Medicine in 1865. Beside this, Pasteur, along with Robert Koch (who was awarded the Nobel Prize in 1905), founded bacteriology. Koch was also famous for the discovery of the tubercle bacillus (1882) and the cholera bacillus (1883) and for his development of Koch's postulates.

The participation of women in medical care (beyond serving as midwives, sitters and cleaning women) was brought about by the likes of Florence Nightingale. These women showed a previously male dominated profession the elemental role of nursing in order to lessen the aggravation of patient mortality which resulted from lack of hygiene and nutrition. Nightingale set up the St Thomas hospital, post-Crimea, in 1852. Elizabeth Blackwell became the first woman to formally study, and subsequently practice, medicine in the United States.

It was in this era that actual cures were developed for certain endemic infectious diseases. However the decline in many of the most lethal diseases was more due to improvements in public health and nutrition than to medicine. It was not until the 20th century that the application of the scientific method to medical research began to produce multiple important developments in medicine, with great advances in pharmacology and surgery.

During the First World War, Alexis Carrel and Henry Dakin developed the Carrel-Dakin method of treating wounds with sutures, which prior to the development of widespread antibiotics, was a major medical progress. The antibiotic prevented the deaths of thousands during the conquest of Vichy France in 1944.

The great war spurred the usage of Roentgen's X-ray, and the electrocardiograph, for the monitoring of internal bodily functions, However, this was overshadowed by the remarkable mass production of penicillum antibiotics, which resulted from government and public pressure.

In the early 1930s scientists in Nazi Germany definitively linked smoking to lung cancer, leading to the most aggressive anti-smoking campaign ever. This knowledge was lost with the 1945 United States' occupation of Germany.

Lunatic asylums began to appear in the Industrial Era. Emil Kraepelin (1856-1926) introduced new medical categories of mental illness, which eventually came into psychiatric usage despite their basis in behavior rather than pathology or etiology. In the 1920s surrealist opposition to psychiatry was expressed in a number of surrealist publications. In the 1930s several controversial medical practices were introduced including inducing seizures (by electroshock, insulin or other drugs) or cutting parts of the brain apart (leucotomy or lobotomy). Both came into widespread use by psychiatry, but there were grave concerns and much opposition on grounds of basic morality, harmful effects, or misuse. In the 1950s new psychiatric drugs, notably the antipsychotic chlorpromazine, were designed in laboratories and slowly came into preferred use. Although often accepted as an advance in some ways, there was some opposition, due to serious adverse effects such as tardive dyskinesia. Patients often opposed psychiatry and refused or stopped taking the drugs when not subject to psychiatric control. There was also increasing opposition to the use of psychiatric hospitals, and attempts to move people back into the community on a collaborative user-led group approach ("therapeutic communities") not controlled by psychiatry. Campaigns against masturbation were done in the Victorian era and elsewhere. Lobotomy was used until the 1970s to treat schizophrenia. This was denounced by the anti-psychiatric movement in the 1960s and later.

The 20th century witnessed a shift from a master-apprentice paradigm of teaching of clinical medicine to a more "democratic" system of medical schools. With the advent of the evidence-based medicine and great advances of information technology the process of change is likely to evolve further, the collation of ideas, resulted in international global projects, such as the Human genome project; However, adversely, the conditions brought about the increasing threat of pandemic spread of mutating diseases, such as SARS, and the danger of the H5N1.

Evidence-based medicine, the application of modern scientific method to ask and answer clinical questions, has had a great impact on practice of medicine throughout the world of modern medicine, for speculation of the unknown was elemental to progress.

Medical Inventions

7000 BC—1000 AD

- c. 7000 BC, drill in Mehrgarh
- c. 7000 BC, bow drill, in Mehrgarh
- c. 7000 BC, dental drill, in Mehrgarh
- c. 7000 BC, surgery, in Mehrgarh
- c. 7000 BC, dental surgery, in Mehrgarh
- c. 2600 BC, suture, by Imhotep
- c. 2600 BC, pharmaceutical cream, by Imhotep
- c. 500 BC, cosmetic surgery, by Sushruta
- c. 500 BC, plastic surgery, by Sushruta
- c. 900 AD, injection syringe, by Ammar ibn Ali al-Mawsili
- 1000, ligature, by Abu al-Qasim (Abulcasis)
- 1000, forceps, by Abu al-Qasim
- 1000, plaster, by Abu al-Qasim
- 1000, curette, by Abu al-Qasim
- 1000, retractor, by Abu al-Qasim

- 1000, scalpel, by Abu al-Qasim
- 1000, sound, by Abu al-Qasim
- 1000, surgical needle, by Abu al-Qasim
- 1000, surgical catgut, by Abu al-Qasim
- 1000, surgical hook, by Abu al-Qasim
- 1000, surgical rod, by Abu al-Qasim
- 1000, surgical spoon, by Abu al-Qasim
- c. 1000, thermometer, by Avicenna
- c. 1000, steam distillation, by Avicenna
- c. 1000, essential oil, by Avicenna

1000—Present

- c. 1280, spectacles
- 1540, artificial limb, by Ambrose Pare
- 1714, mercury thermometer, by Gabriel Fahrenheit
- 1775, bifocal lenses, by Benjamin Franklin
- 1792, ambulance, by Jean Dominique Larrey
- 1796, vaccination, by Edward Jenner
- 1816, stethoscope, by Theophile Laennec
- 1817, dental plate, by Anthony Plantson
- 1827, endoscope, by Pierre Segalas
- 1846, anesthetics, by James Simpson
- 1851, ophthalmoscope, by Hermann von Helmholtz
- 1853, hypodermic syringe, by Alexander Wood
- 1865, antiseptic, by Joseph Lister
- 1885, rabies vaccination, chicken cholera vaccination and by Louis Pasteur
- 1887, contact lens, by Adolf Fick
- 1895, X-ray, by Wilhelm Rontgen
- 1903, electrocardiograph, by Willem Einthoven
- 1905, sphygmomanometer by Nikolai Korotkov
- 1928, penicillin, by Alexander Fleming

- 1938, penicillin as an antibiotic, by Florey and Chain
- 1957, artificial pacemaker, by Clarence Lillehie and Earl Bakken
- 1967, heart transplant, by Christiaan Barnard
- 1973, CAT scan, by Godfrey Hounsfield and Allan Cormack
- 1979, ultrasound scan, by Ian Donald
- 1982, artificial heart, by Robert Jarvik
- 19xx, MRI and fMRI, by

Source

Running Press Cyclopedia, second edition

Special History of Medicine

- History of abortion
- History of alternative medicine
- History of anatomy
- History of brain imaging
- History of cancer chemotherapy
- History of cardiology
- History of invasive and interventional cardiology
- History of endocrinology
- History of immunology
- History of intersex surgery
- History of internal medicine
- History of legal medicine
- History of microbiology
- History of mental illness
- History of neurology
- History of ophthalmology
- History of oto-rhino-laryngology
- History of pharmacology
- History of physiology
- History of psychiatry
- History of surgery

- History of traditional Chinese medicine
- History of veterinary medicine
- History of Islamic medieval ophthalmology
- Timeline of sexual orientation and medicine

Museums and Collections of Health and Medicine

- The London Museums of Health & Medicine
- Osler Library of the History of Medicine
- National Library of Medicine
- Thackray Museum Leeds, in a former workhouse belonging to St James Hospital Thackray Museum

NEUROSCIENCE

Neuroscience is a field that is devoted to the scientific study of the nervous system. Such studies may include the structure, function, development, genetics, biochemistry, physiology, pharmacology, and pathology of the nervous system. Traditionally it is seen as a branch of biological sciences. However, recently there has been a convergence of interest from many allied disciplines, including psychology, computer science, statistics, physics, and medicine. The scope of neuroscience has now broadened to include any systematic scientific experimental and theoretical investigation of the central and peripheral nervous system of biological organisms.

The methodologies employed by neuroscientists have been enormously expanded, from biochemical and genetic analysis of dynamics of individual nerve cells and their molecular constituents to imaging representations of perceptual and motor tasks in the brain.

Neuroscience is at the frontier of investigation of the brain and mind. The study of the brain is becoming the cornerstone in understanding how we perceive and interact with the external world and, in particular, how human experience and human biology influence each other.

The scientific study of the nervous systems underwent a significant increase in the second half of the twentieth century, principally due to revolutions in molecular biology, neural networks and computational neuroscience. It has become possible to understand, in exquisite detail, the complex processes occurring inside a single neuron and in a network that eventually produces the intellectual behavior, cognition, emotion and physiological responses.

The nervous system is composed of a network of neurons and other supportive cells (such as glial cells). Neurons form functional circuits, each responsible for specific tasks to the behaviors at the organism level.

Thus, neuroscience can be studied at many different levels, ranging from molecular level to cellular level to systems level to cognitive level.

At the molecular level, the basic questions addressed in molecular neuroscience include the mechanisms by which neurons express and respond to molecular signals and how axons form complex connectivity patterns. At this level, tools from molecular biology and genetics are used to understand how neurons develop and die, and how genetic changes affect biological functions. The morphology, molecular identity and physiological characteristics of neurons and how they relate to different types of behavior are also of considerable interest. (The ways in which neurons and their connections are modified by experience are addressed at the physiological and cognitive levels.)

At the cellular level, the fundamental questions addressed in cellular neuroscience are the mechanisms of how neurons process signals physiologically and electrochemically. They address how signals are processed by the dendrites, somas and axons, and how neurotransmitters and electrical signals are used to process signals in a neuron.

At the systems level, the questions addressed in systems neuroscience include how the circuits are formed and used anatomically and physiologically to produce the physiological functions, such as reflexes, sensory integration, motor coordination, emotional responses, learning and memory, etc. In other words, they address how these neural circuits function and the mechanisms through which behaviors are generated. For example, systems level analysis addresses questions concerning specific sensory and motor modalities: how does vision work? How do songbirds learn new songs and bats localize with ultrasound? The related field of neuroethology, in particular, addresses the complex question of how neural substrates underlies specific animal behavior.

At the cognitive level, cognitive neuroscience addresses the questions of how psychological/cognitive functions are produced by the neural circuitry. The emergence of powerful experimental techniques such as neuroimaging (*e.g.*,fMRI,PET, SPECT), electrophysiology and human genetic analysis allows neuroscientists to address abstract questions such as how human cognition and emotion are mapped to specific neural circuitries. Many mental processes previously thought to be beyond scientific understanding have been shown to have robust neural correlates.

Neuroscience is also beginning to become allied with social sciences, and burgeoning interdisciplinary fields of neuroeconomics, decision theory, social neuroscience are starting to address some of the most complex questions involving interactions of brain with environment.

Neuroscience generally includes all scientific studies involving the nervous system. Psychology, as the scientific study of mental processes, may be considered a subfield of neuroscience, although some mind/body theorists argue that the definition goes the other way — that psychology is a study of mental processes that can be modeled by many other abstract principles and theories, such as behaviorism and traditional cognitive psychology, that are independent of the underlying neural processes. The term neurobiology is sometimes used interchangeably with neuroscience, though the former refers to the biology of nervous system, whereas the latter refers to science of mental functions that form the foundation of the constituent neural circuitries.

Neurology and Psychiatry are medical specialties and are generally considered, in academic research, subfields of neuroscience that specifically address the diseases of the nervous system. These terms also refer to clinical disciplines involving diagnosis and treatment of these diseases. Neurology deals with diseases of the central and peripheral nervous systems such as amyotrophic lateral sclerosis (ALS) and stroke, while psychiatry focuses on mental illnesses. The boundaries between the two have been blurring recently and physicians who specialize in either generally receive training in both. Both neurology and psychiatry are heavily involved in and influenced by basic research in neuroscience.

History

Evidence of trepanation, the surgical practice of either drilling or scraping a hole into the skull with the aim of curing headaches or mental disorders or relieving cranial pressure, being performed on patients dates back to Neolithic times and has been found in various cultures throughout the world. Manuscripts dating back to 5000BC indicated that the Egyptians had some knowledge about symptoms of brain damage.

Early views on the function of the brain regarded it to be a "cranial stuffing" of sorts. In Egypt, from the late Middle Kingdom onwards, the brain was regularly removed in preparation for mummification. It was believed at the time that the heart was the seat of intelligence. According to Herodotus, during the first step of mummification: 'The most perfect practice is to extract as much of the brain as possible with an iron hook, and what the hook cannot reach is mixed with drugs.'

The view that the heart was the source of consciousness was not challenged until the time of Hippocrates. He believed that the brain was not only involved with sensation, since most specialized organs (*e.g.*, eyes, ears, tongue) are located in the head near the brain, but was also the seat of intelligence. Aristotle, however, believed that the heart was the center of intelligence and that the brain served to cool the blood. This view was generally accepted until the Roman physician Galen, a follower of Hippocrates and physician to Roman gladiators, observed that his patients lost their mental faculties when they had sustained damage to their brains.

In Al-Andalus, Abulcasis, the father of modern surgery, developed material and technical designs which are still used in neurosurgery. Averroes suggested the existence of Parkinson's disease and attributed photoreceptor properties to the retina. Avenzoar described meningitis, intracranial thrombophlebitis, mediastinal tumours and made contributions to modern neuropharmacology. Maimonides wrote about neuropsychiatric disorders and described rabies and belladonna intoxication.

Studies of the brain were became more sophisticated after the invention of the microscope and the development of a staining procedure by Camillo Golgi during the late 1890s that used a silver chromate salt to reveal the intricate structures of single neurons. His technique was used by Santiago Ramón y Cajal and led to the formation of the neuron doctrine, the hypothesis that the functional unit of the brain is the neuron. Golgi and Ramón y Cajal shared the Nobel Prize in Physiology or Medicine in 1906 for their extensive observations, descriptions and categorizations of neurons throughout the brain. The hypotheses of the neuron doctrine were supported by experiments following Galvani's pioneering work in the electrical excitability of muscles and neurons. In the late 19th century, DuBois-Reymond, Müller, and von Helmholtz showed neurons were electrically excitable and that their activity predictably affected the electrical state of adjacent neurons.

In parallel with this research, work with brain-damaged patients by Paul Broca suggested that certain regions of the brain were responsible for certain functions. This hypothesis was supported by observations of epileptic patients conducted by John Hughlings Jackson, who correctly deduced the organization of motor cortex by watching the progression of seizures through the body. Wernicke further developed the theory of the specialization of specific brain structures in language comprehension and production. Modern research still uses the Brodmann cytoarchitectonic (referring to study of cell structure) anatomical definitions from this era in continuing to show that distinct areas of the cortex are activated in the execution of specific tasks.

Main Themes of Research

Neuroscience research from different areas can also be seen as focusing on a set of specific themes and questions. (Some of these are taken from http://www.northwestern.edu/nuin/fac/index.htm)

Brain Mapping

- Behavior/Cognition/Language
- Biological Rhythms
- Brain Imaging or neuroimaging
- Cell Biology
- Cell Imaging & Electrophysiology
- Computational
- Development
- Hearing Sciences
- Learning/Memory
- Mechanisms of Drug Action
- molecular neuroscience
- Motor Control
- Neurobiology of Disease
- Neuroendocrinology
- Neuroimmunology
- Signal transduction
- Systems Neuroscience
- Vision Sciences
- Neurobiology of the neuron
- Sensation and perception
- Sleep
- Autonomic systems and homeostasis
- Arousal, attention and emotion
- Genetics of the nervous system
- Injury of the nervous systems

Neuroscience, by its very interdiciplinary nature, overlaps with and encompasses many different subjects. Bellow is a list of related subjects and fields.

Neuroimaging

Electron Cryomicroscopy (Cryo-EM) is a means of imaging neural serial sections, usually cut by a cryoultramicrotome, now at an experimental maximum (uniform) sample width-reduction of 40-60 nanometers (nm).

A major drawback to this technique is the threat of radiation exposure - and thus biochemical alteration - of the sample's given electron exposure intensity per square unit, time requisite for bioinformatics-retention (radiation exposure time), and the inverse proportionality of high depth-resolution to the material's superficial radition exposure/damage. Also, this is obviously an invasive procedure, meaning the specimen is of a consciousness legally, though not technically dead; for clarification on the definition of death refer to information on cryonics.

Finally, neither serial sectioning (cutting the material - in this case, the brain - into small slices of equal width) nor scanning by electron microscopy are fully automated, often requiring at least partial manual assistance that is tedious and time-consuming to the extreme, making large-scale perception and evaluation impractical given limited funding and general resources. The only exception of this is the nematode, which has the smallest brain known and has been mapped by this method completely.

Application of the attained information is of limited practical significance relative to the shortcomings of any stand-alone mechanism (as well as its resultant metadata complex therefore), and is often optimally beneficial when synthesized in hybrid with the many other imaging techniques, borrowing their appropriate specialized data-perception abilities as compensation for its own short-comings. Most commonly, these complimentary imaging operations include x-ray crystallography (x-ray), magnetic resonance imaging (MRI), scanning electron microscopy, and high-power/magnification light microscopy aided by one or multiple of many available and experimental staining techniques, for this purpose primarily in bath and not by injection (so as to avoid inexorable damage to the neuron so imposed upon, or literally into).

As the first of these four primary choices of examination are non-invasive they are more commonly employed, especially in psychiatric evaluation, which lacks the need present in computational neuroscience (*e.g.* for purposes of virtual neural circuitry as a biological prototype for artificial intelligence (A.I.) development) to image the microscopic mechanisms of, and not simply infer the macro-scale electrical and magnetic properties of the brain. However, electron transmission microscopy is the tool of choice for neuroscientists interested in high-resolution imaging of neural networks and the intracellular biochemical processes that comprise the cornerstone of their morphology, ergo the capability of learning characteristic of intelligence.

- Aphasiology
- Cognitive Science

- **Machine Learning**
- **Neural Networks**
- **Evolutionary neuroscience**
- **Neural engineering**
- **Neuroanatomy**
- **Neurobiology**
- **Neurochemistry**
- **Neuroeconomics**
- **Neuroergonomics**
- **Neuroendocrinology**
- **Neuroesthetics**
- **Neuroethics**
- **Neuroethology**
- **Neurogenetics**
- **Neurogenomics**
- **Neuroheuristic**
- **Neuroimaging**
- **Neurolinguistics**
- **Neuromarketing**
- **Neuropharmacology**
- **Neurophenomenology**
- **Neurophilosophy**
- **Neurophysiology**
- **Neuroproteomics**
- **Neuroprosthetics**
- **Neuropsychiatry**
- **Neuropsychology**
- **Neuropsychopharmacology**
- **Neurotheology (also Biotheology)**
- Psychiatry

- Psychoneuroimmunology
- Psychopharmacology
- Psychobiology (also Biopsychology, also Biological psychology)
- Vision

APPENDIX: UNSOLVED PROBLEMS IN NEUROSCIENCE

Some of the yet unsolved problems of neuroscience include:

- Self awareness: What is the neuronal basis of subjective experience, wakefulness, alertness, arousal and attention? What is its function?
- Perception: How does the brain transfer sensory information into coherent, private percepts? What are the rules by which perception is organized? What are the features/objects that constitute our perceptual experience of internal and external events? How are the senses integrated? Is face perception special (*e.g.* innate)? What is the relationship between subjective experience and the physical world?
- Learning and Memory: Where do our memories get stored and how are they retrieved again? How can learning be improved? What is the difference between explicit and implicit memories? How plastic is the mature brain?
- Development: How and why did the brain evolve (the way it did)? What are the molecular determinants of individual brain development?
- Sleep: Why do we dream? What are the underlying brain mechanisms? What is its relation to anesthesia?
- Cognition and Decisions: How and where does the brain evaluate reward value and effort (cost) to modulate behavior? How does previous experience alter perception and behavior? What are the genetic and environmental contributions to brain function?
- Language: How is it implemented neurally? What is the basis of semantic meaning?
- Diseases: What are the neural bases (causes) of mental diseases like psychotic disorders (*e.g.* mania, schizophrenia), Parkinson's disease, Alzheimer's disease or addiction? Is it possible to recover loss of sensory or motor function?

5

Evolution, Ecology and Evolution Biology

EVOLUTION

In biology, evolution is the change in the inherited traits of a population from one generation to the next. These traits are the expression of genes that are copied and passed on to offspring during reproduction. Mutations in these genes can produce new or altered traits, resulting in heritable differences between organisms. New traits can also come from transfer of genes between populations, as in migration, or between species, in horizontal gene transfer. Evolution occurs when these heritable differences become more common or rare in a population, either non-randomly through natural selection or randomly through genetic drift.

Natural selection is a process that causes *heritable* traits that are helpful for survival and reproduction to become more common, and harmful traits to become more rare. This occurs because organisms with advantageous traits pass on more copies of these heritable traits to the next generation. Over many generations, adaptations occur through a combination of successive, small, random changes in traits, and natural selection of those variants best-suited for their environment. In contrast, genetic drift produces random changes in the frequency of traits in a population. Genetic drift arises from the role chance plays in whether a given individual will survive and reproduce.

One definition of a species is a group of organisms that can reproduce with one another and produce fertile offspring. However, when a species is separated into populations that are prevented from interbreeding, mutations, genetic drift, and the selection of novel traits cause the accumulation of differences over generations and the emergence of new species. The similarities between organisms suggest that all known species are descended from a common ancestor (or ancestral gene pool) through this process of gradual divergence.

The theory of evolution by natural selection was proposed at about the same time by both Charles Darwin and Alfred Russel Wallace, and was set out in detail in Darwin's 1859 book *On the Origin of Species*. It encountered initial resistance from religious authorities who believed humans were divinely set apart from the animal kingdom. In the 1930s, Darwinian natural selection was combined with Mendelian inheritance to form the modern evolutionary synthesis, in which the connection between the *units* of evolution (genes) and the *mechanism* of evolution (natural selection) was made. This powerful explanatory and predictive theory has become the central organizing principle of modern biology, providing a unifying explanation for the diversity of life on Earth.

Heredity

Inheritance in organisms occurs through discrete traits – particular characteristics of an organism. In humans, for example, eye color is an inherited characteristic, which individuals can inherit from one of their parents. Inherited traits are controlled by genes and the complete set of genes within an organism's genome is called its genotype.

The complete set of observable traits that make up the structure and behavior of an organism is called its phenotype. These traits come from the interaction of its genotype with the environment. As a result, not every aspect of an organism's phenotype is inherited. Suntanned skin results from the interaction between a person's genotype and sunlight; thus, a suntan is not hereditary. However, people have different responses to sunlight, arising from differences in their genotype; a striking example is individuals with the inherited trait of albinism, who do not tan and are highly sensitive to sunburn.

Genes are regions within DNA molecules that contain genetic information. DNA is a long molecule with four types of bases attached along its length. Different genes have different sequences of bases; it is the sequence of these bases that encodes genetic information. Within cells, the long strands of DNA associate with proteins to form structures called chromosomes. A specific location within a chromosome is known as a locus. If the DNA sequence at a locus varies between individuals, the different forms of this sequence are called alleles. DNA sequences can change through mutations, producing new alleles. If a mutation occurs within a gene, the new allele may affect the trait that the gene controls, altering the phenotype of the organism. However, while this simple correspondence between an allele and a trait works in some cases, most traits are more complex and are controlled by multiple interacting genes.

Variation

Because an individual's phenotype results from the interaction of their genotype with the environment, the variation in phenotypes in a population reflects the variation in these organisms' genotypes. The modern evolutionary synthesis defines evolution as the change over time in this genetic variation. The frequency of one particular allele will fluctuate, becoming more or less prevalent relative to other forms of that gene. Evolutionary forces act by driving these changes in allele frequency in one direction or another. Variation disappears when an allele reaches the point of fixation — when it either disappears from the population or replaces the ancestral allele entirely.

Variation comes from mutations in genetic material, migration between populations (gene flow), and the reshuffling of genes through sexual reproduction. Variation also comes from exchanges of genes between different species; for example, through horizontal gene transfer in bacteria, and hybridization in plants. Despite the constant introduction of variation through these processes, most of the genome of a species is identical in all individuals of that species. However, even relatively small changes in genotype can lead to dramatic changes in phenotype: chimpanzees and humans differ in only about 5% of their genomes.

Mutation

Genetic variation comes from random mutations that occur in the genomes of organisms. Mutations are changes in the DNA sequence of a cell's genome and are caused by radiation, viruses, transposons and mutagenic chemicals, as well as errors that occur during meiosis or DNA replication. These mutagens produce several different types of change in DNA sequences; these can either have no effect, alter the product of a gene, or prevent the gene from functioning. Studies in the fly *Drosophila melanogaster* suggest that about 70 percent of mutations are deleterious, and the remainder are either neutral or have a weak beneficial effect. Due to the damaging effects that mutations can have on cells, organisms have evolved mechanisms such as DNA repair to remove mutations. Therefore, the optimal mutation rate for a species is a trade-off between short-term costs, such as the risk of cancer, and the long-term benefits of advantageous mutations.

Large sections of DNA can also be duplicated, which is a major source of raw material for evolving new genes, with tens to hundreds of genes duplicated in animal genomes every million years. Most genes belong to larger families of genes of shared ancestry. Novel genes are produced either through duplication and mutation of an ancestral gene, or by recombining parts of different genes to form new combinations with new functions. For example, the human eye uses four genes to make structures that sense light: three for color vision and one for night vision; all four arose from a single ancestral gene. An advantage of duplicating a gene (or even an (entire genome) is that overlapping or redundant functions in multiple genes allows alleles to be retained that would otherwise be harmful, thus increasing genetic diversity.

Changes in chromosome number may also involve the breakage and rearrangement of DNA within chromosomes. For example, two chromosomes in the *Homo* genus fused to produce human chromosome 2; this fusion did not occur in the lineage of the other apes, and they retain these separate chromosomes. The completion of both the human genome project and the chimpanzee genome project has allowed the identification of the relevant chromosomes. In evolution, the most important role of such chromosomal rearrangements may be to accelerate the divergence of a population into new species by preserving genetic differences within populations.

Sequences of DNA that can move about the genome, such as transposons, make up a major fraction of the genetic material of plants and animals, and may have been important in the evolution of genomes. For example, more than a million copies of the Alu sequence are present in the human genome, and these sequences have now been recruited to perform functions such as regulating gene expression. Another effect of these mobile DNA sequences is that when they move within a genome, they can mutate or delete existing genes and thereby produce genetic diversity.

Recombination

In asexual organisms, genes are inherited together, or *linked*, as they cannot mix with genes in other organisms during reproduction. However, the offspring of sexual organisms contain a random mixture of their parents' chromosomes that is produced through independent assortment. In the related process of genetic recombination, sexual organisms can also exchange DNA between two matching chromosomes. These shuffling processes can allow even alleles that are close together in a strand of DNA to be inherited independently. However, as only about one recombination event occurs per million base pairs in humans, genes close together on a chromosome may not be shuffled away from each other, and tend to be inherited together. This tendency is measured by finding how often two alleles occur together, which is called their linkage disequilibrium. A set of alleles that is usually inherited in a group is called a haplotype, and this co-inheritance can indicate that the locus is under positive selection.

Recombination in sexual organisms helps to remove harmful mutations and retain beneficial mutations. Consequently, when alleles cannot be separated by recombination – such as in mammalian Y chromosomes, which pass intact from fathers to sons – harmful mutations accumulate. In addition, recombination can produce individuals with new and advantageous gene combinations. These positive effects of recombination are balanced by the fact that this process can cause mutations and separate beneficial combinations of genes. The optimal rate of recombination for a species is therefore a trade-off between conflicting factors.

Mechanisms

There are three basic mechanisms of evolutionary change: natural selection, genetic drift, and gene flow. Natural selection favors genes that improve capacity for survival and reproduction. Genetic drift is random change in the frequency of alleles, caused by the random sampling of a generation's genes during reproduction, and gene flow is the transfer of genes within and between populations. The relative importance of natural selection and genetic drift in a population varies depending on the strength of the selection and the effective population size, which is the number of individuals capable of breeding. Natural selection usually predominates in large populations, while genetic drift dominates in small populations. The dominance of genetic drift in small populations can even lead to the fixation of slightly deleterious mutations. As a result, changing population size can dramatically influence the course of evolution. Population bottlenecks, where the population shrinks temporarily and therefore loses genetic variation, result in a more uniform population. Bottlenecks also result from alterations in gene flow such as decreased migration, expansions into new habitats, or population subdivision.

Natural Selection

Natural selection is the process by which genetic mutations that enhance reproduction become, and remain, more common in successive generations of a population. It has often been called a "self-evident" mechanism because it necessarily follows from three simple facts:

- Heritable variation exists within populations of organisms.
- Organisms produce more offspring than can survive.
- These offspring vary in their ability to survive and reproduce.

These conditions produce competition between organisms for survival and reproduction. Consequently, organisms with traits that give them an advantage over their competitors pass these advantageous traits on, while traits that do not confer an advantage are not passed on to the next generation.

The central concept of natural selection is the evolutionary fitness of an organism.

This measures the organism's genetic contribution to the next generation. However, this is not the same as the total number of offspring: instead fitness measures the proportion of subsequent generations that carry an organism's genes. Consequently, if an allele increases fitness more than the other alleles of that gene, then with each generation this allele will become more common within the population. These traits are said to be "selected *for*". Examples of traits that can increase fitness are enhanced survival, and increased fecundity. Conversely, the lower fitness caused by having a less beneficial or deleterious allele results in this allele becoming rarer — they are "selected *against*". Importantly, the fitness of an allele is not a fixed characteristic, if the environment changes, previously neutral or harmful traits may become beneficial and previously beneficial traits become harmful.

Natural selection within a population for a trait that can vary across a range of values, such as height, can be categorized into three different types. The first is directional selection, which is a shift in the average value of a trait over time — for example organisms slowly getting taller. Secondly, disruptive selection is selection for extreme trait values and often results in two different values becoming most common, with selection against the average value. This would be when either short or tall organisms had an advantage, but not those of medium height. Finally, in stabilizing selection there is selection against extreme trait values on both ends, which causes a decrease in variance around the average value. This would, for example, cause organisms to slowly become all the same height.

A special case of natural selection is sexual selection, which is selection for any trait that increases mating success by increasing the attractiveness of an organism to potential mates. Traits that evolved through sexual selection are particularly prominent in males of some animal species, despite traits such as cumbersome antlers, mating calls or bright colors that attract predators, decreasing the survival of individual males. This survival disadvantage is balanced by higher reproductive success in males that show these hard to fake, sexually selected traits.

An active area of research is the unit of selection, with natural selection being proposed to work at the level of genes, cells, individual organisms, groups of organisms and even species. None of these models are mutually-exclusive and selection may act on multiple levels simultaneously. Below the level of the individual, genes called transposons try to copy themselves throughout the genome. Selection at a level above the individual, such as group selection, may allow the evolution of co-operation, as discussed below.

Genetic Drift

Genetic drift is the change in allele frequency from one generation to the next that occurs because alleles in the offspring generation are a random sample of those in the parent generation, and are thus subject to sampling error. As a result, when selective forces are absent or relatively weak, allele frequencies tend to "drift" upward or downward in a random walk. This drift halts when an allele eventually becomes fixed, either by disappearing from the population, or replacing the other alleles entirely. Genetic drift may therefore eliminate some alleles from a population due to chance alone, and two separate populations that began with the same genetic structure can drift apart by random fluctuation into two divergent populations with different sets of alleles. The time for an allele to become fixed by genetic drift depends on population size, with fixation occurring more rapidly in smaller populations.

Although natural selection is responsible for adaptation, the relative importance of the two forces of natural selection and genetic drift in driving evolutionary change in general is an area of current research in evolutionary biology. These investigations were prompted by the neutral theory of molecular evolution, which proposed that most evolutionary changes are the result the fixation of neutral mutations that do not have any immediate effects on the fitness of an organism. Hence, in this model, most genetic changes in a population are the result of constant mutation pressure and genetic drift.

Gene Flow

Gene flow is the exchange of genes between populations, which are usually of the same species. Examples of gene flow within a species include the migration and then breeding of organisms, or the exchange of pollen. Gene transfer between species includes the formation of hybrid organisms and horizontal gene transfer.

Migration into or out of a population can change allele frequencies. Immigration may add new genetic material to the established gene pool of a population. Conversely, emigration may remove genetic material. As barriers to reproduction between two diverging populations are required for the populations to become new species, gene flow may slow this process by spreading genetic differences between the populations. Gene flow is hindered by mountain ranges, oceans and deserts or even man-made structures such as the Great Wall of China, which has hindered the flow of plant genes.

Depending on how far two species have diverged since their most recent common ancestor, it may still be possible for them to produce offspring, as with horses and donkeys mating to produce mules. Such hybrids are generally infertile, due to the two different sets of chromosomes being unable to pair up during meiosis. In this case, closely-related species may regularly interbreed, but hybrids will be selected against and the species will remain distinct. However, viable hybrids are occasionally formed and these new species can either have properties intermediate between their parent species, or possess a totally new phenotype. The importance of hybridization in creating new species of animals is unclear, although cases have been seen in many types of animals, with the gray tree frog particularly well-studied.

Hybridization is, however, an important means of speciation in plants, since polyploidy (having more than two copies of each chromosome) is tolerated in plants more readily than in animals. Polyploidy is important in hybrids as it allows reproduction, with the two different sets of chromosomes each being able to pair with an identical partner during meiosis. Polyploids also have more genetic diversity, which allows them to avoid inbreeding depression in small populations.

Horizontal gene transfer is the transfer of genetic material from one organism to another organism that is not its offspring, this is most common among bacteria. In medicine, this contributes to the spread of antibiotic resistance, as when one bacteria acquires resistance genes it can rapidly transfer them to other species. Horizontal transfer of genes from bacteria to eukaryotes such as the yeast *Saccharomyces cerevisiae* and the adzuki bean beetle *Callosobruchus chinensis* may also have occurred. Viruses can also carry DNA between organisms, allowing transfer of genes even across biological domains. Gene transfer has also occurred within eukaryotic cells, from the chloroplast and mitochondrial genomes to nuclear genomes.

Outcomes

Evolution influences every aspect of the form and behavior of organisms. Most prominent are the specific behavioral and physical adaptations that are the outcome of natural selection. These adaptations increase fitness by aiding activities such as finding food, avoiding predators or attracting mates. Organisms can also respond to selection by co-operating with each other, usually by aiding their relatives or engaging in mutually-beneficial symbiosis. In the longer term, evolution produces new species through splitting ancestral populations of organisms into new groups that are unable to breed with one another.

These outcomes of evolution are sometimes divided into macroevolution, which is evolution that occurs at or above the level of species, such as speciation, and microevolution, which is smaller evolutionary changes, such as adaptations, within a species or population. In general, macroevolution is the outcome of long periods of microevolution. Thus, the distinction between micro- and macroevolution is not a fundamental one - the difference is simply the time involved. However, in macroevolution, the traits of the entire species are important. For instance, a large amount of variation among individuals allows a species to rapidly adapt to new habitats, lessening the chance of it going extinct, while a wide geographic range increases the chance of speciation, by making it more likely that part of the population will become isolated. In this sense, microevolution and macroevolution can sometimes be separate.

A common misconception is that evolution is "progressive," but natural selection has no long-term goal and does not necessarily produce greater complexity. Although complex species have evolved, this occurs as a side effect of the overall number of organisms increasing, and simple forms of life remain more common. For example, the overwhelming majority of species are microscopic prokaryotes, which form about half the world's biomass despite their small size, and constitute the vast majority of Earth's biodiversity. Simple organisms therefore remain the dominant form of life on Earth, and complex life appears more diverse only because it is more noticeable.

Adaptation

Adaptations are structures or behaviors that enhance a specific function, causing organisms to become better at surviving and reproducing. They are produced by a combination of the continuous production of small, random changes in traits, followed by natural selection of the variants best-suited for their environment. This process can cause either the gain of a new feature, or the loss of an ancestral feature. An example that shows both types of change is bacterial adaptation to antibiotic selection, with mutations causing antibiotic resistance by either modifying the target of the drug, or removing the transporters that allow the drug into the cell. However, many traits that appear to be simple adaptations are in fact exaptations: structures originally adapted for one function, but which coincidentally became somewhat useful for some other function in the process. One example is the African lizard *Holapsis guentheri*, which developed an extremely flat head for hiding in crevices, as can be seen by looking at its near relatives. However, in this species, the head has become so flattened that it assists in gliding from tree to tree - an exaptation.

As adaptation occurs through the gradual modification of existing structures, structures with similar internal organization may have very different functions in related organisms. This is the result of a single ancestral structure being adapted to function in different ways. The bones within bat wings, for example, are structurally similar to both human hands and seal flippers, due to the common descent of these structures from an ancestor that also had five digits at the end of each forelimb. Other idiosyncratic anatomical features, such as bones in the wrist of the panda being formed into a false "thumb," indicate that an organism's evolutionary lineage can limit what adaptations are possible. During adaption, some structures may lose their original function and become vestigial structures. Such structures may have little or no function in a current species, yet have a clear function in ancestral species, or other closely-related species. Examples include the non-functional remains of eyes in blind cave-dwelling fish, wings in flightless birds, and the presence of hip bones in whales and snakes. Examples of vestigial structures in humans include wisdom teeth, the coccyx, and the vermiform appendix.

An area of current investigation in evolutionary developmental biology is the developmental basis of adaptations and exaptations. This research addresses the origin and evolution of embryonic development and how modifications of development and developmental processes produce novel features. These studies have shown that evolution can alter development to create new structures, such as embryonic bone structures that develop into the jaw in other animals instead forming part of the middle ear in mammals. It is also possible for structures that have been lost in evolution to reappear due to changes in developmental genes, such as a mutation in chickens causing embryos to grow teeth similar to those of crocodiles.

Co-Evolution

Interactions between organisms can produce both conflict and co-operation. When the interaction is between pairs of species, such as a pathogen and a host, or a predator and its prey, these species can develop matched sets of adaptations. Here, the evolution of one species causes adaptations in a second species. These changes in the second species then, in turn, cause new adaptations in the first species. This cycle of selection and response is called co-evolution. An example is the production of tetrodotoxin in the rough-skinned newt and the evolution of tetrodotoxin resistance in its predator, the common garter snake. In this predator-prey pair, an evolutionary arms race has produced high levels of toxin in the newt and correspondingly high levels of resistance in the snake.

Co-Operation

However, not all interactions between species involve conflict. Many cases of mutually beneficial interactions have evolved. For instance, an extreme cooperation exists between plants and the mycorrhizal fungi that grow on their roots and aid the plant in absorbing nutrients from the soil. This is a reciprocal relationship as the plants provide the fungi with sugars from photosynthesis. Here, the fungi actually grow inside plant cells, allowing them to exchange nutrients with their hosts, while sending signals that suppress the plant immune system.

Coalitions between organisms of the same species have also evolved. An extreme case is the Eusociality found in social insects, such as bees, termites and ants, where sterile insects feed and guard the small number of organisms in a colony that are able to reproduce. On an even smaller scale, the somatic cells that make up the body of an animal are limited in their capacity to reproduce in order to maintain a stable organism which then supports a small number of the animal's germ cells to produce offspring. Here, somatic cells respond to specific signals that instruct them to either grow or kill themselves. If cells ignore these signals and attempt to multiply inappropriately, their uncontrolled growth causes cancer.

These examples of cooperation within species are thought to have evolved through the process of kin selection, which is where one organism acts to help raise a relative's offspring. This activity is selected for because if the *helping* individual contains alleles which promote the helping activity, it is likely that its kin will *also* contain these alleles and thus those alleles will be passed on. Other processes that may promote cooperation include group selection, where cooperation provides benefits to a group of organisms.

Speciation

Speciation is the process where a species diverges into two or more descendant species. It has been observed multiple times under both controlled laboratory conditions and in nature. In sexually-reproducing organisms, speciation results from reproductive isolation followed by genealogical divergence. There are four mechanisms for speciation. The most common in animals is allopatric speciation, which occurs in populations initially isolated geographically, such as by habitat fragmentation or migration. As selection and drift act independently in isolated populations, separation will eventually produce organisms that cannot interbreed.

The second mechanism of speciation is peripatric speciation, which occurs when small populations of organisms become isolated in a new environment. This differs from allopatric speciation in that the isolated populations are numerically much smaller than the parental population. Here, the founder effect causes rapid speciation through both rapid genetic drift and selection on a small gene pool.

The third mechanism of speciation is parapatric speciation. This is similar to peripatric speciation in that a small population enters a new habitat, but differs in that there is no physical separation between these two populations. Instead, speciation results from the evolution of mechanisms that reduce gene flow between the two populations. Generally this occurs when there has been a drastic change in the environment within the parental species' habitat. One example is the grass *Anthoxanthum odoratum*, which can undergo parapatric speciation in response to localized metal pollution from mines. Here, plants evolve that have resistance to high levels of metals in the soil. Selection against interbreeding with the metal-sensitive parental population produces a change in flowering time of the metal-resistant plants, causing reproductive isolation. Selection against hybrids between the two populations may cause *reinforcement*, which is the evolution of traits that promote mating within a species, as well as character displacement, which is when two species become more distinct in appearance.

Finally, in sympatric speciation species diverge without geographic isolation or changes in habitat. This form is rare since even a small amount of gene flow may remove genetic differences between parts of a population. Generally, sympatric speciation in animals requires the evolution of both genetic differences and non-random mating, to allow reproductive isolation to evolve.

One type of sympatric speciation involves cross-breeding of two related species to produce a new hybrid species. This is not common in animals as animal hybrids are usually sterile, because during meiosis the homologous chromosomes from each parent, being from different species cannot successfully pair. It is more common in plants, however because plants often double their number of chromosomes, to form polyploids. This allows the chromosomes from each parental species to form a matching pair during meiosis, as each parent's chromosomes is represented by a pair already. An example of such a speciation event is when the plant *Oenothera lamarckiana* gave rise to a new species of plant *O. gigas* in 1905, which had a different chromosome number than its parents. (O. gigas had a chromosome number of 2N = 28 while its parent *Oenothera lamarckiana* had a chromosome number of 2N = 14.)

Indeed, chromosome doubling can itself cause reproductive isolation, as half the doubled chromosomes will be unmatched when breeding with undoubled organisms.

Speciation events are important in the theory of punctuated equilibrium, which accounts for the pattern in the fossil record of short "bursts" of evolution interspersed with relatively long periods of stasis, where species remain relatively unchanged. In this theory, speciation and rapid evolution are linked, with natural selection and genetic drift acting most strongly on organisms undergoing speciation in novel habitats or small populations. As a result, the periods of stasis in the fossil record correspond to the parental population, and the organisms undergoing speciation and rapid evolution are found in small populations or geographically-restricted habitats, and therefore rarely being preserved as fossils.

Extinction

Extinction is the disappearance of an entire species. Extinction is not an unusual event, as species regularly appear through speciation, and disappear through extinction. Indeed, virtually all animal and plant species that have lived on earth are now extinct. These extinctions have happened continuously throughout the history of life, although the rate of extinction spikes in occasional mass extinction events. The Cretaceous–Tertiary extinction event, during which the dinosaurs went extinct, is the most well-known, but the earlier Permian-Triassic extinction event was even more severe, with approximately 96 percent of species driven to extinction. The Holocene extinction event is an ongoing mass extinction associated with humanity's expansion across the globe over the past few thousand years. Present-day extinction rates are 100-1000 times greater than the background rate, and up to 30 percent of species may be extinct by the mid 21st century. Human activities are now the primary cause of the ongoing extinction event; global warming may further accelerate it in the

The role of extinction in evolution depends on which type is considered. The causes of the continuous "low-level" extinction events, which form the majority of extinctions, are not well understood and may be the result of competition between species for shared resources. If competition from other species does alter the probability that a species will become extinct, this could produce species selection as a level of natural selection. The intermittent mass extinctions are also important, but instead of acting as a selective force, they drastically reduce diversity in a nonspecific manner and promote bursts of rapid evolution and speciation in survivors.

Evolutionary History of Life

Origin of Life

The origin of life is a necessary precursor for biological evolution, but understanding that evolution occurred once organisms appeared and investigating how this happens, does not depend on understanding exactly how life began. The current scientific consensus is that the complex biochemistry that makes up life came from simpler chemical reactions, but it is unclear how this occurred. Not much is certain about the earliest developments in life, the structure of the first living things, or the identity and nature of any last universal common ancestor or ancestral gene pool. Consequently, there is no scientific consensus on how life began, but proposals include self-replicating molecules such as RNA, and the assembly of simple cells.

Common Descent

All organisms on Earth are descended from a common ancestor or ancestral gene pool. Current species are a stage in the process of evolution, with their diversity the product of a long series of speciation and extinction events. The common descent of organisms was first deduced from four simple facts about organisms: First, they have geographic distributions that cannot be explained by local adaptation. Second, the diversity of life is not a set of completely unique organisms, but organisms that share morphological similarities. Third, vestigial traits with no clear purpose resemble functional ancestral traits, and finally, that organisms can be classified using these similarities into a hierarchy of nested groups.

Past species have also left records of their evolutionary history. Fossils, along with the comparative anatomy of present-day organisms, constitute the morphological, or anatomical, record. By comparing the anatomies of both modern and extinct species, paleontologists can infer the lineages of those species. However, this approach is most successful for organisms that had hard body parts, such as shells, bones or teeth. Further, as prokaryotes such as bacteria and archaea share a limited set of common morphologies, their fossils do not provide information on their ancestry.

More recently, evidence for common descent has come from the study of biochemical similarities between organisms. For example, all living cells use the same nucleic acids and amino acids. The development of molecular genetics has revealed the record of evolution left in organisms' genomes: dating when species diverged through the molecular clock produced by mutations. For example, these DNA sequence comparisons have revealed the close genetic similarity between humans and chimpanzees and shed light on when the common ancestor of these species existed.

Evolution of Life

Despite the uncertainty on how life began, it is clear that prokaryotes were the first organisms to inhabit Earth, approximately 3–4 billion years ago. No obvious changes in morphology or cellular organization occurred in these organisms over the next few billion years.

The eukaryotes were the next major innovation in evolution. These came from ancient bacteria being engulfed by the ancestors of eukaryotic cells, in a cooperative association called endosymbiosis. The engulfed bacteria and the host cell then underwent co-evolution, with the bacteria evolving into either mitochondria or hydrogenosomes. An independent second engulfment of cyanobacterial-like organisms led to the formation of chloroplasts in algae and plants.

The history of life was that of the unicellular eukaryotes, prokaryotes, and archaea until about a billion years ago when multicellular organisms began to appear in the oceans in the Ediacaran period. The evolution of multicellularity occurred in multiple independent events, in organisms as diverse as sponges, brown algae, cyanobacteria, slime moulds and myxobacteria.

Soon after the emergence of these first multicellular organisms, a remarkable amount of biological diversity appeared over approximately 10 million years, in an event called the Cambrian explosion. Here, the majority of types of modern animals evolved, as well as unique lineages that subsequently became extinct. Various triggers for the Cambrian explosion have been proposed, including the accumulation of oxygen in the atmosphere from photosynthesis. About 500 million years ago, plants and fungi colonized the land, and were soon followed by arthropods and other animals. Amphibians first appeared around 300 million years ago, followed by early amniotes, then mammals around 200 million years ago and birds around 100 million years ago (both from "reptile"-like lineages). However, despite the evolution of these large animals, smaller organisms similar to the types that evolved early in this process continue to be highly successful and dominate the Earth, with the majority of both biomass and species being prokaryotes.

History of Evolutionary Thought

Evolutionary ideas such as common descent and the transmutation of species have existed since at least the 6th century BC, when they were expounded by the Greek philosopher Anaximander. Evolutionary thought was further developed by other early thinkers, including the Greek philosopher Empedocles, the Roman philosopher Lucretius, the Arab biologist Al-Jahiz, the Persian philosopher Ibn Miskawayh, and the Brethren of Purity. Also in the Far East, the philosopher Zhuangzi discussed a transformative power of species to adapt to their surroundings. As biological knowledge grew in the 18th century, a variety of such ideas developed, beginning with Pierre Maupertuis in 1745, and with contributions from natural philosophers such as Erasmus Darwin and Jean-Baptiste Lamarck. In 1858, Charles Darwin and Alfred Russel Wallace jointly proposed the theory of evolution by natural selection to the Linnean Society of London in separate papers. Shortly after, Darwin's publication of *On the Origin of Species* provided detailed support for the theory and led to increasingly wide acceptance of the occurrence of evolution.

Nonetheless, Darwin's specific ideas about evolution, such as gradualism and the mechanisms of natural selection, were strongly contested at first. Lamarckists argued that transmutation of species occurred as parents passed on adaptations acquired during their lifetimes. Eventually, when experiments failed to support it, this idea was abandoned in favor of Darwinism. More significantly, Darwin could not account for how traits were passed down from generation to generation. A mechanism was provided in 1865 by Gregor Mendel, who found that traits were inherited in a predictable manner. When Mendel's work was rediscovered in 1900, disagreements over the rate of evolution predicted by early geneticists and biometricians led to a rift between the Mendelian and Darwinian models of evolution.

This contradiction was reconciled in the 1930s by biologists such as Ronald Fisher. The end result was a combination of evolution by natural selection and Mendelian inheritance, the modern evolutionary synthesis. In the 1940s, the identification of DNA as the genetic material by Oswald Avery and colleagues and the subsequent publication of the structure of DNA by James Watson and Francis Crick in 1953, demonstrated the physical basis for inheritance. Since then, genetics and molecular biology have become core parts of evolutionary biology and have revolutionized the field of phylogenetics.

In its early history, evolutionary biology primarily drew in scientists from traditional taxonomically-oriented disciplines, whose specialist training in particular organisms addressed general questions in evolution. As evolutionary biology expanded as an academic discipline, particularly after the development of the modern evolutionary synthesis, it began to draw more widely from the biological sciences. Currently the study of evolutionary biology involves scientists from fields as diverse as biochemistry, ecology, genetics and physiology, and evolutionary concepts are used in even more distant disciplines such as psychology, medicine, philosophy and computer science.

Social and Cultural Views

Even before the publication of *On the Origin of Species*, the idea that life had evolved was a source of debate and evolution is still a contentious concept. Debate has generally centered on the philosophical, social and religious implications of evolution, not on the science itself; the proposition that biological evolution occurs through the mechanism of natural selection is standard in the scientific literature.

As Darwin recognized early on, the most controversial aspect of evolutionary thought is its implications to humans. Specifically, some people believe that the origin of humans involved supernatural intervention, rather than natural processes. Although many religions and denominations have reconciled their beliefs with evolution through theistic evolution, several denominations contain creationists who object to evolution, as it contradicts their literal interpretation of origin beliefs. In some countries – notably the United States – these tensions between scientific and religious teachings have fueled the ongoing creation–evolution controversy, a religious conflict focusing on politics and public education. While other scientific fields such as cosmology and earth science also conflict with literal interpretations of many religious texts, evolutionary biology has strong oppositions from believers of religion.

Evolution has been used to support philosophical positions that promote discrimination and racism. For example, the eugenic ideas of Francis Galton were developed to argue that the human gene pool should be improved by selective breeding policies, including incentives for those considered "good stock" to reproduce, and the compulsory sterilization, prenatal testing, birth control, and even killing, of those considered *bad stock*. Another example of an extension of evolutionary theory that is now widely regarded as unwarranted is "Social Darwinism," a term given to the 19th century Whig Malthusian theory developed by Herbert Spencer into ideas about "survival of the fittest" in commerce and human societies as a whole, and by others into claims that social inequality, racism, and imperialism were justified. However, contemporary scientists and philosophers consider these ideas to have been neither mandated by evolutionary theory nor supported by data.

Applications in Technology

A major technological application of evolution is artificial selection, which is the intentional selection of certain traits in a population of organisms. Humans have used artificial selection for thousands of years in the domestication of plants and animals. More recently, such selection has become a vital part of genetic engineering, with selectable markers such as antibiotic resistance genes being used to manipulate DNA in molecular biology.

As evolution can produce highly optimized processes and networks, it has many applications in computer science. Here, simulations of evolution using evolutionary algorithms and artificial life started with the work of Nils Aall Barricelli in the 1960s, and was extended by Alex Fraser, who published a series of papers on simulation of artificial selection. Artificial evolution became a widely recognized optimization method as a result of the work of Ingo Rechenberg in the 1960s and early 1970s, who used evolution strategies to solve complex engineering problems. Genetic algorithms in particular became popular through the writing of John Holland. As academic interest grew, dramatic increases in the power of computers allowed practical applications. Evolutionary algorithms are now used to solve multi-dimensional problems more efficiently than software produced by human designers, and also to optimize the design of systems.

ECOLOGY AND EVOLUTION BIOLOGY

I. ECOLOGY

Ecology is the scientific study of the distribution and abundance of living organisms and how the distribution and abundance are affected by interactions between the organisms and their environment. The environment of an organism includes both physical properties, which can be described as the sum of local abiotic factors such as insolation (sunlight), climate, and geology, and biotic factors, which are other organisms that share its habitat. The term *oekologie* was coined in 1866 by the German biologist Ernst Haeckel.

The word "ecology" is often used in common parlance as a synonym for the natural environment or environmentalism. Likewise "ecologic" or "ecological" is often taken in the sense of environmentally friendly.

Ecology is usually considered a branch of biology, the general science that studies living organisms. Organisms can be studied at many different levels, from proteins and nucleic acids (in biochemistry and molecular biology), to cells (in cellular biology), to individuals (in botany, zoology, and other similar disciplines), and finally at the level of populations, communities, and ecosystems, to the biosphere as a whole; these latter strata are the primary. subjects of ecological inquiries.

Ecology is a multi-disciplinary science. Because of its focus on the higher levels of the organization of life on earth and on the interrelations between organisms and their environment, ecology draws heavily on many other branches of science, especially geology and geography, meteorology, pedology, genetics, chemistry, and physics. Thus, ecology is considered by some to be a holistic science, one that over-arches older disciplines such as biology which in this view become sub-disciplines contributing to ecological knowledge.

Agriculture, fisheries, forestry, medicine and urban development are among human activities that would fall within Krebs' (1972: 4) explanation of his definition of ecology: *where organisms are found, how many occur there, and why.*

As a scientific discipline, ecology does not dictate what is "right" or "wrong". However, ecological knowledge such as the quantification of biodiversity and population dynamics have provided a scientific basis for expressing the aims of environmentalism and evaluating its goals and policies. Additionally, a holistic view of nature is stressed in both ecology and environmentalism.

Consider the ways an ecologist might approach studying the life of honeybees:

- The behavioral relationship between individuals of a species is behavioral ecology — for example, the study of the queen bee, and how she relates to the worker bees and the drones.
- The organized activity of a species is community ecology; for example, the activity of bees assures the pollination of flowering plants. Bee hives additionally produce honey which is consumed by still other species, such as bears.
- The relationship between the environment and a species is environmental ecology — for example, the consequences of environmental change on bee activity. Bees may die out due to environmental changes. The environment simultaneously affects and is a consequence of this activity and is thus intertwined with the survival of the species.

Disciplines

Ecology is a broad discipline comprising many sub-disciplines. A common, broad classification, moving from lowest to highest complexity, where complexity is defined as the number of entities and processes in the system under study, is:

- Ecophysiology and Behavioral ecology examine adaptations of the individual to its environment.
- Autecology studies the dynamics of populations of a single species.

- Community ecology (or synecology) focuses on the interactions between species within an ecological community.
- Ecosystem ecology studies the flows of energy and matter through the biotic and abiotic components of ecosystems.
- Landscape ecology examines processes and relationship across multiple ecosystems or very large geographic areas.

Ecology can also be sub-divided according to the species of interest into fields such as animal ecology, plant ecology, insect ecology, and so on. Another frequent method of subdivision is by biome studied, *e.g.*, Arctic ecology (or polar ecology), tropical ecology, desert ecology, etc. The primary technique used for investigation is often used to subdivide the discipline into groups such as chemical ecology, genetic ecology, field ecology, statistical ecology, theoretical ecology, and so forth. These fields are not mutually exclusive; one could be a theoretical plant community ecologist, or a polar ecologist interested in animal genetics. Animals can be reproduced by plants.

Fundamental Principles

Biosphere

For modern ecologists, ecology can be studied at several levels: population level (individuals of the same species in the same or similar environment), biocenosis level (or community of species), ecosystem level, and biosphere level.

The outer layer of the planet Earth can be divided into several compartments: the hydrosphere (or sphere of water), the lithosphere (or sphere of soils and rocks), and the atmosphere (or sphere of the air). The biosphere (or sphere of life), sometimes described as "the fourth envelope", is all living matter on the planet or that portion of the planet occupied by life. It reaches well into the other three spheres, although there are no permanent inhabitants of the atmosphere. Relative to the volume of the Earth, the biosphere is only the very thin surface layer which extends from 11,000 meters below sea level to 15,000 meters above.

It is thought that life first developed in the hydrosphere, at shallow depths, in the photic zone. (Although recently a competing theory has emerged, that life originated around hydrothermal vents in the deeper ocean. See Origin of life.) Multicellular organisms then appeared and colonized benthic zones. Photosynthetic organisms gradually produced the chemically unstable oxygen-rich atmosphere that characterizes our planet. Terrestrial life developed later, after the ozone layer protecting living beings from UV rays formed. Diversification of terrestrial species is thought to be increased by the continents drifting apart, or alternately, colliding. Biodiversity is expressed at the ecological level (ecosystem), population level (intraspecific diversity), species level (specific diversity), and genetic level. Recently technology has allowed the discovery of the deep ocean vent communities. This remarkable ecological system is not dependent on sunlight but bacteria, utilising the chemistry of the hot volcanic vents, are at the base of its food chain.

The biosphere contains great quantities of elements such as carbon, nitrogen and oxygen. Other elements, such as phosphorus, calcium, and potassium, are also essential to life, yet are present in smaller amounts. At the ecosystem and biosphere levels, there is a continual recycling of all these elements, which alternate between the mineral and organic states.

While there is a slight input of geothermal energy, the bulk of the functioning of the ecosystem is based on the input of solar energy. Plants and photosynthetic microorganisms convert light into chemical energy by the process of photosynthesis, which creates glucose (a simple sugar) and releases free oxygen. Glucose thus becomes the secondary energy source which drives the ecosystem. Some of this glucose is used directly by other organisms for energy. Other sugar molecules can be converted to other molecules such as amino acids. Plants use some of this sugar, concentrated in nectar to entice pollinators to aid them in reproduction.

Cellular respiration is the process by which organisms (like mammals) break the glucose back down into its constituents, water and carbon dioxide, thus regaining the stored energy the sun originally gave to the plants. The proportion of photosynthetic activity of plants and other photosynthesizers to the respiration of other organisms determines the specific composition of the Earth's atmosphere, particularly its oxygen level. Global air currents mix the atmosphere and maintain nearly the same balance of elements in areas of intense biological activity and areas of slight biological activity.

Water is also exchanged between the hydrosphere, lithosphere, atmosphere and biosphere in regular cycles. The oceans are large tanks, which store water, ensure thermal and climatic stability, as well as the transport of chemical elements thanks to large oceanic currents.

For a better understanding of how the biosphere works, and various dysfunctions related to human activity, American scientists simulated the biosphere in a small-scale model, called Biosphere II.

The Ecosystem Concept

The first principle of ecology is that each living organism has an ongoing and continual relationship with every other element that makes up its environment. An ecosystem can be defined as any situation where there is interaction between organisms and their environment.

The ecosystem is composed of two entities, the entirety of life, the biocoenosis and the medium that life exists in, the biotope. Within the ecosystem, species are connected by food chains or food webs. Energy from the sun, captured by primary producers via photosynthesis, flows upward through the chain to primary consumers (herbivores), and then to secondary and tertiary consumers (carnivores), before ultimately being lost to the system as waste heat. In the process, matter is incorporated into living organisms, which return their nutrients to the system via decomposition, forming biogeochemical cycles such as the carbon and nitrogen cycles.

The concept of an ecosystem can apply to units of variable size, such as a pond, a field, or a piece of dead wood. An ecosystem within another ecosystem is called a micro ecosystem. *For example, an ecosystem can be a stone and all the life under it.* A meso ecosystem *could be a forest, and a* macro ecosystem *a whole eco region, with its drainage basin.*

The main questions when studying an ecosystem are:

- Whether the colonization of a barren area could be carried out
- Investigation the ecosystem's dynamics and changes
- The methods of which an ecosystem interacts at local, regional and global scale
- Whether the current state is stable
- Investigating the value of an ecosystem and the ways and means that interaction of ecological systems provides benefits to humans, especially in the provision of healthy water.

Ecosystems are often classified by reference to the biotopes concerned. The following ecosystems may be defined:

- As continental ecosystems, such as forest ecosystems, meadow ecosystems such as steppes or savannas, or agro-ecosystems
- As ecosystems of inland waters, such as lentic ecosystems such as lakes or ponds; or lotic ecosystems such as rivers
- As oceanic ecosystems.

Another classification can be done by reference to its communities, such as in the case of an human ecosystem.

Spatial Relationships and Subdivisions of Land

Ecosystems are not isolated from each other, but are interrelated. For example, water may circulate between ecosystems by the means of a river or ocean current.

Water itself, as a liquid medium, even defines ecosystems. Some species, such as salmon or freshwater eels move between marine systems and fresh-water systems. These relationships between the ecosystems lead to the concept of a *biome*.

A biome is a homogeneous ecological formation that exists over a large region as tundra or steppes. The biosphere comprises all of the Earth's biomes — the entirety of places where life is possible — from the highest mountains to the depths of the oceans.

Biomes correspond rather well to subdivisions distributed along the latitudes, from the equator towards the poles, with differences based on to the physical environment (for example, oceans or mountain ranges) and to the climate. Their variation is generally related to the distribution of species according to their ability to tolerate temperature and/or dryness. For example, one may find photosynthetic algae only in the *photic* part of the ocean (where light penetrates), while conifers are mostly found in mountains.

Though this is a simplification of more complicated scheme, latitude and altitude approximate a good representation of the distribution of biodiversity within the biosphere. Very generally, the richness of biodiversity (as well for animal than plant species) is decreasing most rapidly near the equator and less rapidly as one approaches the poles. The biosphere may also be divided into ecozones, which are very well defined today and primarily follow the continental borders. The ecozones are themselves divided into ecoregions, though there is not agreement on their limits.

Dynamics and Stability

Ecological factors which affect dynamic change in a population or species in a given ecology or environment are usually divided into two groups: abiotic and biotic.

Abiotic factors are geological, geographical, hydrological and climatological parameters. A biotope is an environmentally uniform region characterized by a particular set of abiotic ecological factors. Specific abiotic factors include:

- Water, which is at the same time an essential element to life and a milieu
- Air, which provides oxygen, nitrogen, and carbon dioxide to living species and allows the dissemination of pollen and spores
- Soil, at the same time source of nutriment and physical support - Soil pH, salinity, nitrogen and phosphorus content, ability to retain water, and density are all influential
- Temperature, which should not exceed certain extremes, even if tolerance to heat is significant for some species

- Light, which provides energy to the ecosystem through photosynthesis
- Natural disasters can also be considered abiotic

Biocenose, or community, is a group of populations of plants, animals, micro-organisms. Each population is the result of procreations between individuals of same species and cohabitation in a given place and for a given time. When a population consists of an insufficient number of individuals, that population is threatened with extinction; the extinction of a species can approach when all biocenoses composed of individuals of the species are in decline. In small populations, consanguinity (inbreeding) can result in reduced genetic diversity that can further weaken the biocenose.

Biotic ecological factors also influence biocenose viability; these factors are considered as either intraspecific and interspecific relations. An ant lion lies in wait under its pit trap, built in dry dust under a building, awaiting unwary insects that fall in. Many pest insects are partly or wholly controlled by other insect predators. Interspecific relations—interactions between different species—are numerous, and usually described according to their beneficial, detrimental or neutral effect (for example, mutualism (relation ++) or competition (relation —). The most significant relation is the relation of predation (to eat or to be eaten), which leads to the essential concepts in ecology of food chains (for example, the grass is consumed by the herbivore, itself consumed by a carnivore, itself consumed by a carnivore of larger size). A high predator to prey ratio can have a negative influence on both the predator and prey biocenoses in that low availability of food and high death rate prior to sexual maturity can decrease (or prevent the increase of) populations of each, respectively. Selective hunting of species by humans which leads to population decline is one example of a high predator to prey ratio in action. Other interspecific relations include parasitism, infectious disease and competition for limiting resources, which can occur when two species share the same ecological niche.

The existing interactions between the various living beings go along with a permanent mixing of mineral and organic substances, absorbed by organisms for their growth, their maintenance and their reproduction, to be finally rejected as waste. These permanent recyclings of the elements (in particular carbon, oxygen and nitrogen) as well as the water are called biogeochemical cycles. They guarantee a durable stability of the biosphere (at least when unchecked human influence and extreme weather or geological phenomena are left aside). This self-regulation, supported by negative feedback controls, ensures the perenniality of the ecosystems. It is shown by the very stable concentrations of most elements of each compartment. This is referred to as homeostasis. The ecosystem also tends to evolve to a state of ideal balance, reached after a succession of events, the climax (for example a pond can become a peat bog).

Ecological Crisis

Generally, an ecological crisis occurs with the loss of adaptive capacity when the resilience of an environment or of a species or a population evolves in a way unfavourable to coping with perturbations that interfere with that ecosystem, landscape or species survival.

It may be that the environment quality degrades compared to the species needs, after a change in an abiotic ecological factor (for example, an increase of temperature, less significant rainfalls).

It may be that the environment becomes unfavourable for the survival of a species (or a population) due to an increased pressure of predation (for example overfishing). Lastly, it may be that the situation becomes unfavourable to the quality of life of the species (or the population) due to a rise in the number of individuals (overpopulation).

Ecological crises may be more or less brutal (occurring within a few months or taking as long as a few million years). They can also be of natural or anthropic origin. They may relate to one unique species or to many species.

Lastly, an ecological crisis may be local (as an oil spill) or global (a rise in the sea level due to global warming).

According to its degree of endemism, a local crisis will have more or less significant consequences, from the death of many individuals to the total extinction of a species. Whatever its origin, disappearance of one or several species often will involve a rupture in the food chain, further impacting the survival of other species.

In the case of a global crisis, the consequences can be much more significant; some extinction events showed the disappearance of more than 90% of existing species at that time. However, it should be noted that the disappearance of certain species, such as the dinosaurs, by freeing an ecological niche, allowed the development and the diversification of the mammals. An ecological crisis thus paradoxically favored biodiversity.

Sometimes, an ecological crisis can be a specific and reversible phenomenon at the ecosystem scale. But more generally, the crises impact will last. Indeed, it rather is a connected series of events, that occur till a final point. From this stage, no return to the previous stable state is possible, and a new stable state will be set up gradually.

Lastly, if an ecological crisis can cause extinction, it can also more simply reduce the quality of life of the remaining individuals. Thus, even if the diversity of the human population is sometimes considered threatened, few people envision human disappearance at short span. However, epidemic diseases, famines, impact on health of reduction of air quality, food crises, reduction of living space, accumulation of toxic or non degradable wastes, threats on keystone species (great apes, panda, whales) are also factors influencing the well-being of people.

During the past decades, this increasing responsibility of humanity in some ecological crises has been clearly observed. Due to the increases in technology and a rapidly increasing population, humans have more influence on their own environment than any other ecosystem engineer.

Some usually quoted examples as ecological crises are:

- Permian-Triassic extinction event 250 million of years ago
- Cretaceous-Tertiary extinction event 65 million years ago
- Global warming related to the Greenhouse effect. Warming could involve flooding of the Asian deltas, multiplication of extreme weather phenomena and changes in the nature and quantity of the food resource.
- Ozone layer hole issue
- Deforestation and desertification, with disappearance of many species.
- The nuclear meltdown at Chernobyl in 1986 caused the death of many people and animals from cancer, and caused mutations in a large number of animals and people. The area around the plant is now abandoned by humans because of the large amount of radiation generated by the meltdown. Twenty years after the accident, the animals have returned.

Ecosystem Productivity

In an ecosystem, the connections between species are generally related to food and their role in the food chain. There are three categories of organisms:

- *Decomposers* — bacteria, mushrooms which degrade organic matter of all categories, and restore minerals to the environment.
- *Consumers* — animals, which can be primary consumers (herbivorous), or secondary or tertiary consumers (carnivorous).
- *Producers* — usually plants which are capable of photosynthesis but could be other organisms such as bacteria around ocean vents that are capable of chemosynthesis.

These relations form sequences, in which each individual consumes the preceding one and is consumed by the one following, in what are called food chains or food network. In a food network, there will be fewer organisms at each level as one follows the links of the network up the chain. These concepts lead to the idea of biomass (the total living matter in a given place), of primary productivity (the increase in the mass of plants during a given time) and of secondary productivity (the living matter produced by consumers and the decomposers in a given time).

These two last ideas are key, since they make it possible to evaluate the load capacity — the number of organisms which can be supported by a given ecosystem. In any food network, the energy contained in the level of the producers is not completely transferred to the consumers. And the higher one goes up the chain, the more energy and resources is lost and consumed. Thus, from an energy—and environmental—point of view, it is more efficient for humans to be primary consumers (to subsist from vegetables, grains, legumes, fruit, etc.) than as secondary consumers (from eating herbivores, omnivores, or their products, such as milk, chickens, cattle, sheep, etc.) and still more so than as a tertiary consumer (from consuming carnivores, omnivores, or their products, such as fur, pigs, snakes, alligators, etc.). An ecosystem(s) is unstable when the load capacity is overrun and is especially unstable when a population doesn't have an ecological niche and overconsumers. The productivity of ecosystems is sometimes estimated by comparing three types of land-based ecosystems and the total of aquatic ecosystems:

- The forests (1/3 of the Earth's land area) contain dense biomasses and are very productive. The total production of the world's forests corresponds to half of the primary production.
- Extreme ecosystems in the areas with more extreme climates — deserts and semi-deserts, tundra, alpine meadows, and steppes — (1/3 of the Earth's land area) have very sparse biomasses and low productivity
- Savannas, meadows, and marshes (1/3 of the Earth's land area) contain less dense biomasses, but are productive. These ecosystems represent the major part of what humans depend on for food.
- Finally, the marine and fresh water ecosystems (3/4 of Earth's surface) contain very sparse biomasses (apart from the coastal zones).

Humanity's actions over the last few centuries have seriously reduced the amount of the Earth covered by forests (deforestation), and have increased agro-ecosystems (agriculture). In recent decades, an increase in the areas occupied by extreme ecosystems has occurred (desertification).

II. Evolutionary Biology

Evolutionary biology is a sub-field of biology concerned with the origin and descent of species, as well as their change, multiplication, and diversity over time. One who studies evolutionary biology is known as an evolutionary biologist, or less formally, an evolutionist.

Evolutionary biology is an interdisciplinary field because it includes scientists from a wide range of both field and lab oriented disciplines. For example, it generally includes scientists who may have a specialist training in particular organisms such as mammalogy, ornithology, or herpetology, but use those organisms as case studies to answer general questions in evolution. It also generally includes paleontologists and geologists who use fossils to answer questions about the tempo and mode of evolution, as well as theoreticians in areas such as population genetics and evolutionary psychology. In the 1990s developmental biology made a re-entry into evolutionary biology from its initial exclusion from the modern synthesis through the study of evolutionary developmental biology.

Its findings feed strongly into new disciplines that study mankind's sociocultural evolution and evolutionary behavior. Evolutionary biology's frameworks of ideas and conceptual tools are now finding application in the study of a range of subjects from computing to nanotechnology. Artificial life is a sub-field of Bioinformatics that attempts to model, or even recreate, the evolution of organisms as described by evolutionary biology. Usually this is done through mathematics and computer models.

Evolutionary biology as an academic discipline in its own right emerged as a result of the modern evolutionary synthesis in the 1930s and 1940s. It was not until the 1970s and 1980s, however, that a significant number of universities had departments that specifically included the term *evolutionary biology* in their titles. In the United States, as a result of the rapid growth of molecular and cell biology, many universities have split (or aggregated) their biology departments into *molecular and cell biology*-style departments and *ecology and evolutionary biology*-style departments (which often have subsumed older departments in paleontology, zoology and the like).

Microbiology has recently developed into an evolutionary discipline. It was originally ignored due to the paucity of morphological traits and the lack of a species concept in microbiology. Now, evolutionary researchers are taking advantage our extensive understanding of microbial physiology, the ease of microbial genomics, and the quick generation time of some microbes to answer evolutionary questions. Similar features have led to progress in viral evolution, particularly for bacteriophage.

6

Introduction to Technology and Bioscience Technology

TECHNOLOGY

Technology is a broad concept that deals with a species' usage and knowledge of tools and crafts, and how it affects a species' ability to control and adapt to its environment. In human society, it is a consequence of science and engineering, although several technological advances predate the two concepts. However, a strict definition is elusive; "technology" can refer to material objects of use to humanity, such as machines, hardware or utensils, but can also encompass broader themes, including systems, methods of organization, and techniques. The term can either be applied generally or to specific areas: examples include "construction technology", "medical technology", or "state-of-the-art technology".

The human race's use of technology began with the conversion of natural resources into simple tools. The prehistorical discovery of the ability to control fire increased the available sources of food and the invention of the wheel helped humans in travelling in and controlling their environment. Recent technological developments, including the printing press, the telephone, and the Internet, have lessened physical barriers to communication and allowed humans to interact on a global scale. However, not all technology has been used for peaceful purposes; the development of weapons of ever-increasing destructive power has progressed throughout history, from clubs to nuclear weapons.

Technology has affected society and its surroundings in a number of ways. In many societies, technology has helped develop more advanced economies (including today's global economy) and has allowed the rise of a leisure class. Many technological

processes produce unwanted by-products, known as pollution, and deplete natural resources, to the detriment of the Earth and its environment. Various implementations of technology influence the values of a society and new technology often raises new ethical questions. Examples include the rise of the notion of efficiency in terms of human productivity, a term originally applied only to machines, and the challenge of traditional norms.

Philosophical debates have arisen over the present and future use of technology in society, with disagreements over whether technology improves the human condition or worsens it. Neo-Luddism, anarcho-primitivism, and similar movements criticise the pervasiveness of technology in the modern world, claiming that it harms the environment and alienates people; proponents of ideologies such as transhumanism and techno-progressivism view continued technological progress as beneficial to society and the human condition. Indeed, until recently, it was believed that the development of technology was restricted only to human beings, but recent scientific studies indicate that other primates and certain dolphin communities have developed simple tools and learned to pass their knowledge to other generations.

Definition and Usage

In general technology is the relationship that society has with its tools and crafts, and to what extent society can control its environment.

- The Merriam-Webster dictionary offers a definition of the term: "the practical application of knowledge especially in a particular area" and "a capability given by the practical application of knowledge".
- Ursula Franklin, in her 1989 "Real World of Technology" lecture, gave another definition of the concept; it is "practice, the way we do things around here". The term is often used to imply a specific field of technology, or to refer to high technology, rather than technology as a whole.
- Bernard Stiegler, in *Technics and Time, 1*, defines technology in two ways: as "the pursuit of life by means other than life," and as "organized inorganic matter."

The term is mostly used in three different contexts: when referring to a tool (or machine); a technique; the cultural force; or a combination of the three.

Technology can be most broadly defined as the entities, both material and immaterial, created by the application of mental and physical effort in order to achieve some value. In this usage, technology refers to tools and machines that may be used to solve real-world problems. It is a far-reaching term that may include simple tools, such as a crowbar or wooden spoon, or more complex machines, such as a space station or particle accelerator. Tools and machines need not be material;

virtual technology, such as computer software and business methods, fall under this definition of technology.

The word "technology" can also be used to refer to a collection of techniques. In this context, it is the current state of humanity's knowledge of how to combine resources to produce desired products, to solve problems, fulfill needs, or satisfy wants; it includes technical methods, skills, processes, techniques, tools and raw materials. When combined with another term, such as "medical technology" or "space technology", it refers to the state of the respective field's knowledge and tools. "State-of-the-art technology" refers to the high technology available to humanity in any field.

Technology can be viewed as an activity that forms or changes culture. Additionally, technology is the application of math, science, and the arts for the benefit of life as it is known. A modern example is the rise of communication technology, which has lessened barriers to human interaction and, as a result, has helped spawn new subcultures; the rise of cyberculture has, at its basis, the development of the Internet and the computer. Not all technology enhances culture in a creative way; technology can also help facilitate political oppression and war via tools such as guns. As a cultural activity, technology predates both science and engineering, each of which formalize some aspects of technological endeavor.

Science, Engineering and Technology

The distinction between science, engineering and technology is not always clear. Science is the reasoned investigation or study of phenomena, aimed at discovering enduring principles among elements of the phenomenal world by employing formal techniques such as the scientific method. Technologies are not usually exclusively products of science, because they have to satisfy requirements such as utility, usability and safety.

Engineering is the goal-oriented process of designing and making tools and systems to exploit natural phenomena for practical human means, often (but not always) using results and techniques from science. The development of technology may draw upon many fields of knowledge, including scientific, engineering, mathematical, linguistic, and historical knowledge, to achieve some practical result.

Technology is often a consequence of science and engineering — although technology as a human activity preceeds the two fields. For example, science might study the flow of electrons in electrical conductors, by using already-existing tools and knowledge. This new-found knowledge may then be used by engineers to create new tools and machines, such as semiconductors, computers, and other forms of advanced technology. In this sense, scientists and engineers may both be considered technologists; the three fields are often considered as one for the purposes of research and reference.

History

Prehistory (—500BCE)

The history of technology is at least as old as humankind, if not older. Primitive tools have been discovered with almost every find of ancient human remains. Archaeologists have uncovered tools made by humanity's ancestors more than two million years ago, and the earliest direct evidence of tool usage, found in the Great Rift Valley, dates back to 2.5 million years ago. The hunter-gatherer lifestyle, characteristic of the Lower Paleolithic era, involved a limited use of technology, and the earliest tools, such as the handaxe and scraper, were developed to aid early humans in that role.

The discovery and utilization of fire, a simple energy source with many profound uses, was a turning point in the technological evolution of humankind. The exact date of its discovery is not known; evidence of burnt animal bones at the Cradle of Humankind suggests that the domestication of fire occurred before 1,000,000 BCE; scholarly consensus indicates that Homo erectus had controlled fire by between 500,000 BCE and 400,000 BCE. Fire, fueled with wood and charcoal, allowed early humans to cook their food to increase its digestibility, improving its nutrient value and broadening the number of foods that could be eaten.

Other technological advances made during the Paleolithic era were clothing and shelter; the adoption of both technologies cannot be dated exactly, but they were key to humanity's progress. As the Paleolithic era progressed, dwellings became more sophisticated and more elaborate; as early as 380,000 BCE, humans were constructing temporary wood huts. Clothing, adapted from the fur and hides of hunted animals, helped humanity expand into colder regions; humans began to migrate out of Africa by 200,000 BCE and into other continents, such as Eurasia.

A more sophisticated toolmaking technique was developed at around the same time. Known as the prepared-core technique, it enabled the creation of more controlled and consistent flakes, which could be hafted onto wooden shafts as arrows. This new technique helped to form more efficient composite tools and weapons, and combined with fire, this new technique enabled humans to hunt more effectively; wooden spears with fire-hardened points have been found as early as 250,000 BCE.

Technological developments in the Upper Paleolithic era, helped by the development of language, included advances in flint tool manufacturing, with industries based on fine blades rather than simple flakes. Humans began to work bones, antler, and hides, as evidenced by burins and racloirs produced during this period.

Ancient History (5000BCE — 0CE)

Continuing improvements led to the furnace and bellows and provided the ability to smelt and forge native metals (naturally occurring in relatively pure form). Gold, copper, silver, and lead, were such early metals. The advantages of copper tools over stone, bone, and wooden tools were quickly apparent to early humans, and native copper was probably used from near the beginning of Neolithic times (about 8000 BCE). Native copper does not naturally occur in large amounts, but copper ores are quite common and some of them produce metal easily when burned in wood or charcoal fires. Eventually, the working of metals led to the discovery of alloys such as bronze and brass (about 4000 BCE). The first uses of iron alloys such as steel dates to around 1400 BCE.

Meanwhile, humans were learning to harness other forms of energy. The earliest known use of wind power is the sailboat. The earliest record of a ship under sail is shown on an Egyptian pot dating back to 3200 BCE. From prehistoric times, Egyptians probably used "the power of the Nile" annual floods to irrigate their lands, gradually learning to regulate much of it through purposely-built irrigation channels and 'catch' basins. Similarly, the early peoples of Mesopotamia, the Sumerians, learned to use the Tigris and Euphrates rivers for much the same purposes. But more extensive use of wind and water (and even human) power required another invention.

According to archaeologists, the wheel was invented around 4000 B.C. The wheel was likely independently invented in Mesopotamia (in present-day Iraq) as well. Estimates on when this may have occurred range from 5500 to 3000 B.C., with most experts putting it closer to 4000 B.C. The oldest artifacts with drawings that depict wheeled carts date from about 3000 B.C.; however, the wheel may have been in use for millenia before these drawings were made. There is also evidence from the same period of time that wheels were used for the production of pottery. (Note that the original potter's wheel was probably not a wheel, but rather an irregularly shaped slab of flat wood with a small hollowed or pierced area near the center and mounted on a peg driven into the earth. It would have been rotated by repeated tugs by the potter or his assistant.) More recently, the oldest-known wooden wheel in the world was found in the Ljubljana marshes of Slovenia.

The invention of the wheel revolutionized activities as disparate as transportation, war, and the production of pottery (for which it may have been first used). It didn't take long to discover that wheeled wagons could be used to carry heavy loads and fast (rotary) potters' wheels enabled early mass production of pottery. But it was the use of the wheel as a transformer of energy (through water wheels, windmills, and even treadmills) that revolutionized the application of nonhuman power sources.

Modern History (0CE —)

Tools include both simple machines (such as the lever, the screw, and the pulley), and more complex machines (such as the clock, the engine, the electric generator and the electric motor, the computer, radio, and the Space Station, among many others).

As tools increase in complexity, so does the type of knowledge needed to support them. Complex modern machines require libraries of written technical manuals of collected information that has continually increased and improved — their designers, builders, maintainers, and users often require the mastery of decades of sophisticated general and specific training. Moreover, these tools have become so complex that a comprehensive infrastructure of technical knowledge-based lesser tools, processes and practices (complex tools in themselves) exist to support them, including engineering, medicine, and computer science. Complex manufacturing and construction techniques and organizations are needed to construct and maintain them. Entire industries have arisen to support and develop succeeding generations of increasingly more complex tools.

Technology and Society

The relationship of technology with society (and/or culture) is generally characterized as synergistic, symbiotic, co-dependent, co-influential, and co-producing, i.e. technology and society depend heavily one upon the other (technology upon culture, and culture upon technology). It is also generally believed that this synergistic relationship first occurred at the dawn of humankind with the invention of simple tools, and continues with modern technologies today. Today and throughout history, technology influences and is influenced by such societal issues/factors as economics, values, ethics, institutions, groups, the environment, government, among others.

The discipline studying the impacts of science, technology, and society and vice versa is called (and can be found at) Science and technology studies.

Technology and Philosophy

Technicism

Generally, technicism is an over reliance or overconfidence in technology as a benefactor of society.

Taken to extreme, some argue that technicism is the belief that humanity will ultimately be able to control the entirety of existence using technology. In other words, human beings will eventually be able to master all problems, supply all wants and needs, possibly even control the future. Some, such as Monsma, connect these ideas to the abdication of religion as a higher moral authority.

More commonly, technicism is a criticism of the commonly held belief that newer, more recently-developed technology is "better." For example, more recently-developed computers are faster than older computers, and more recently-developed cars have greater gas efficiency and more features than older cars. Because current technologies are generally accepted as good, future technological developments are not considered circumspectly, resulting in what seems to be a blind acceptance of technological developments.

Optimism

Optimistic assumptions are made by proponents of ideologies such as transhumanism and singularitarianism, which view technological development as generally having beneficial effects for the society and the human condition. In these ideologies, technological development is morally good. Some critics see these ideologies as examples of scientism and techno-utopianism and fear the notion of human enhancement and technological singularity which they support. Some have described Karl Marx as a techno-optimist.

Pessimism

On the somewhat pessimistic side are certain philosophers like the Herbert Marcuse and John Zerzan, who believe that technological societies are inherently flawed *a priori*. They suggest that the result of such a society is to become evermore technological at the cost of freedom and psychological health (and probably physical health in general, as pollution from technological products is dispersed).

Many, such as the Luddites and prominent philosopher Martin Heidegger, hold serious reservations, although not *a priori* flawed reservations, about technology. Wrote Heidegger in "The Question Concerning Technology": "Thus we shall never experience our relationship to the essence of technology so long as we merely conceive and push forward the technological, put up with it, or evade it. Everywhere we remain unfree and chained to technology, whether we passionately affirm or deny it."

Some of the most poignant criticisms of technology are found in what are now considered to be dystopian literary classics, for example Aldous Huxley's *Brave New World* and other writings, Anthony Burgess's *A Clockwork Orange*, and George Orwell's *Nineteen Eighty-Four*. And, in *Faust* by Goethe, Faust's selling his soul to the devil in return for power over the physical world, is also often interpreted as a metaphor for the adoption of industrial technology.

An overtly anti-technological treatise is *Industrial Society and Its Future*, written by Theodore Kaczynski (aka The Unabomber) and printed in several major newspapers (and later books) as part of an effort to end his bombing campaign of the techno-industrial infrastructure.

Appropriate Technology

The notion of appropriate technology, however, was developed in the 20th century (e.g., see the work of Jacques Ellul) to describe situations where it was not desirable to use very new technologies or those that required access to some centralized infrastructure or parts or skills imported from elsewhere. The eco-village movement emerged in part due to this concern.

Other Species

The use of basic technology is also a feature of other species apart from humans. These include primates such as chimpanzees, some dolphin communities, and crows.

The ability to make and use tools was once considered a defining characteristic of the genus Homo. However, the discovery of tool construction among chimpanzees and related primates has discarded the notion of the use of technology as unique to humans. For example, researchers have observed wild chimpanzees utilising tools for foraging: some of the tools used include leaf sponges, termite fishing probes, pestles and levers. West African chimpanzees also use stone hammers and anvils for cracking nuts.

HISTORY OF TECHNOLOGY

The history of technology is the history of the invention of tools and techniques for background knowledge has enabled us to create new things, and conversely, many scientific endeavors have become possible through technologies which assist humans to travel to places we could not otherwise go, and probe the nature of the universe in more detail than our natural senses allow.

Technological artifacts are products of an economy, a force for economic growth, and a large part of everyday life. Technological innovations affect, and are affected by, a society's cultural traditions. They also are a means to develop and project military power.

By Period and Geography

Early Technology

- Olduvai stone technology (Olduwan) 2.5 million years ago (scrapers; to butcher dead animals)
- Acheulean stone technology 1.6 million years ago (hand axe)
- Fire used since the Paleolithic, possibly by Homo erectus as early as 1.5 Million years ago

- **Clothing possibly 100,000 years ago.**
- **Stone tools, used by Homo floresiensis, possibly 100,000 years ago.**
- **Domestication of Animals, ca. 15,000 BC**
- **Pottery ca. 11th millennium BC**
- **Bow, sling ca. 9th millennium BC**
- **Microliths ca. 9th millennium BC**
- **Copper ca. 8000 BC**
- **Agriculture and Plough ca. 8000 BC**
- **Wheel ca. 4000 BC**
- **Gnomon ca. 4000 BC**
- **Writing systems ca. 3500 BC**
- **Bronze ca. 3300 BC**
- **Salt**
- **Chariot ca. 2000 BC**
- **Iron ca. 1500 BC**
- **Sundial ca. 800 BC**
- **Glass ca. 500 BC**
- **Catapult ca. 400 BC**
- **Horseshoe ca. 300 BC**
- **Stirrup first few centuries AD**

Stone Age

During the Stone Age, all humans were hunter-gatherers, a lifestyle which involved limited use of tools and few if any permanent settlements. The first major technologies, then, were tied to survival, hunting, and food preparation in this environment. Fire, stone tools and weapons, and clothing were technological developments of major importance during this period. Stone Age cultures developed music, and engaged in organized warfare. A subset of Stone Age people developed ocean-worthy outrigger ship technology, leading to an eastward migration across the Malay archipelago, across the Indian ocean to Madagascar and also across the Pacific Ocean, which required knowledge of the ocean currents, weather patterns, sailing, celestial navigation, and star maps. The early Stone Age is described as Epipaleolithic or Mesolithic. The former is generally used to describe the early Stone Age in areas with limited glacial impact. The later Stone Age, during which

the rudiments of agricultural technology were developed, is called the Neolithic period.

Although Paleolithic cultures left no written records, the shift from nomadic life to settlement and agriculture can be inferred from a range of archaeological evidence. Such evidence includes ancient tools, cave paintings, and other prehistoric art, such as the Venus of Willendorf. Human remains also provide direct evidence, both through the examination of bones, and the study of mummies. Though concrete evidence is limited, scientists and historians have been able to form significant inferences about the lifestyle and culture of various prehistoric peoples, and the role technology played in their lives.

Copper and Bronze Age

The Stone Age developed into the Bronze Age after the Neolithic Revolution. The Neolithic Revolution involved radical changes in agricultural technology which included development of agriculture, animal domestication, and the adoption of permanent settlements. These combined factors made possible the development of metal smelting, with copper and later bronze, being the metals of choice. This technological trend began in the Fertile Crescent, and spread outward over time. It should be noted that these developments were not, and still are not, universal. The Three-age system does not accurately describe the technology history of groups outside of Eurasia, and does not apply at all in the case of some isolated populations, such as the Spinifex People, the Sentinelese, and various Amazonian tribes, which still make use of Stone Age technology, and have not developed agricultural or metallurgical technology...

Iron Age

The Iron Age involved the adoption of iron smelting technology. It generally replaced bronze, and made it possible to produce tools which were stronger and cheaper to make than bronze equivalents. In many Eurasian cultures, the Iron Age was the last major step before the development of written language, though again this was not universally the case.

Ancient Civilizations

Egypt

The Egyptians invented and used many simple machines, such as the ramp and the lever, to aid construction processes. Egyptian paper, made from papyrus, and pottery was mass produced and exported throughout the Mediterranean basin. The wheel, however, did not arrive until foreign invaders introduced the chariot. They also played an important role in developing Mediterranean maritime technology including ships and lighthouses.

For later technologies in Ptolemaic Egypt, Roman Egypt, and Arab Egypt, see Ancient Greek technology and innovation, Roman technology and Inventions in the Muslim world respectively.

India

The Indus Valley Civilization, situated in a resource-rich area, is notable for its early application of city planning and sanitation technologies. Cites in the Indus Valley offer some of the first examples of closed gutters, public baths, and communal granaries. The Takshashila University was an important seat of learning in the ancient world. It was the center of education for scholars from all over Asia. Many Greek, Persian and Chinese students studied here under great scholars including Kautilya, Panini, Jivaka, and Vishnu Sharma.

Ancient India was also at the forefront of seafaring technology - a panel found at Mohenjodaro, depicts a sailing craft. Ship construction is vividly described in the Yukti Kalpa Taru, an ancient Indian text on Shipbuilding. The Yukti Kalpa Taru, compiled by Bhoja Narapati is concerned with shipbuilding.

Indian construction and architecture, called 'Vaastu Shastra', suggests a thorough understanding or materials engineering, hydrology, and sanitation. Ancient Indian culture was also pioneering in its use of vegetable dyes, cultivating plants including indigo and cinnabar. Many of the dyes were used in art and sculpture. The use of perfumes demonstrates some knowledge of chemistry, particularly distillation and purification processes.

China

According to the Scottish researcher Joseph Needham, the Chinese made a great many first-known discoveries and developments. Major technological contributions from China include early seismological detectors, matches, paper, sliding calipers, the double-action piston pump, cast iron, the iron plough, the multi-tube seed drill, the wheelbarrow, the suspension bridge, the parachute, natural gas as fuel, the magnetic compass, the raised-relief map, the propeller, the crossbow, the South Pointing Chariot, and gun powder. Other Chinese discoveries and inventions from the Medieval period, according to Joseph Needham's research, include: the paddle wheel boat, block printing and movable type, phosphorescent paint, chain drive, the escapement mechanism, and the spinning wheel.

The solid-fuel rocket was invented in China about 1150, nearly 200 years after the invention of black powder (which acted as the rocket's fuel), and 500 years after the invention of the match. At the same time that the age of exploration was occurring in the West, the Chinese emperors of the Ming Dynasty also sent ships, some reaching Africa. But the enterprises were not further funded, halting further exploration and development. When Ferdinand Magellan's ships reached Brunei in 1521, they found a wealthy city that had been fortified by Chinese engineers, and protected by a breakwater. Antonio Pigafetta noted that much of the technology of Brunei was equal to Western technology of the time. Also, there were more cannons in Brunei than on Magellan's ships, and the Chinese merchants to the Brunei court had sold them spectacles and porcelain, which were rarities in Europe. Chinese scientific understanding, however, was less developed than that in the West.

Tribal Europe

By 1000 BC - 500 BC, the Germanic tribes had a Bronze Age civilization, while the Celts were in the Iron Age by the time of the Hallstatt culture. Their cultures collided with the military and agricultural practices of the Romans, leading the Europeans to appropriate both social and technological processes of the Romans.

Greek and Hellenistic

Greek and Hellenistic engineers invented many technologies and improved upon pre-existing technologies, particularly during the Hellenistic period. Heron of Alexandria invented a basic steam engine and demonstrated knowledge of mechanic and pneumatic systems. Archimedes invented several machines. The Greeks were unique in pre-industrial times in their ability to combine scientific research with the development of new technologies. One example is the Archimedean screw; this technology was first conceptualized in mathematics, then built. Other technologies invented by Greek scientists include the ballistae, and primitive analog computers like the Antikythera mechanism and the piston pump. Greek architects were responsible for the first true domes, and were the first to explore the Golden ratio and its relationship with geometry and architecture.

Apart from Hero of Alexandria's steam aeolipile, Hellenistic technicians were the first to invent watermills and windwheels, making them global pioneers in three of the four known means of non-human propulsion prior to the Industrial Revolution (the fourth being sails), although only water power became extensively used in antiquity.

Roman

Romans developed an intensive and sophisticated agriculture, expanded upon existing iron working technology, created laws providing for individual ownership, advanced stone masonry technology, advanced road-building (exceeded only in the 19th century), military engineering, civil engineering, spinning and weaving and several different machines like the Gallic reaper that helped to increase productivity in many sectors of the Roman economy.

Roman engineers were the first to build monumental arches, amphitheatres, aqueducts, public baths, stone bridges, vaults and domes. Notable Roman inventions include the book (Codex), glass blowing and concrete. Because Rome was located on a volcanic peninsula, with sand which contained suitable crystalline grains, the concrete which the Romans formulated was especially durable. Some of their buildings have lasted 2000 years, to the present day.

Roman civilization was highly urbanized by pre-modern standards. Many cities of the Imperium had over 100,000 inhabitants with the capital Rome being the largest metropolis of antiquity. Features of Roman urban life included multistory apartment buildings, street paving, public flush toilets, glass windows and floor and wall heating. The Romans understood hydraulics and constructed fountains and waterworks, particularly Aqueducts, which were the hallmark of their civilization. Some Roman baths have lasted to this day. The Romans developed many technologies which were lost in the Middle Ages, and were only fully reinvented in the 19th and 20th centuries.

Inca

The engineering skills of the Inca were great, even by today's standards. An example is the use of pieces weighing in upwards of one ton in their stonework (e.g., Machu Picchu in Peru), placed together so that not even a blade can fit in-between the cracks. The villages used irrigation canals and drainage systems, making agriculture very efficient. While some claim that the Incas were the first inventors of hydroponics, their agricultural technology was still soil based, if advanced. This technology, including tiered farm plots, allowed significant yields from steeply sloped or otherwise unproductive land.

Maya

Though the Maya civilization had no metallurgy or wheel technology, they developed complex writing and astrological systems, and created sculptural works in stone and flint. Like the Inca, the Maya also had command of fairly advanced agricultural and construction technology. Throughout this time period much of this construction, was made only by women, as men of the Maya civilization believed that females were responsible for the creation of new things.

Medieval and Modern Technologies

Muslim Agricultural Revolution

From the 8th century, the medieval Islamic world witnessed a fundamental transformation in agriculture known as the "Muslim Agricultural Revolution", "Arab Agricultural Revolution", or "Green Revolution". Due to the global economy established by Muslim traders across the Old World during the "Afro-Asiatic age of discovery" or "Pax Islamica", this enabled the diffusion of many crops, plants and farming techniques between different parts of the Islamic world, as well as the adaptation of crops, plants and techniques from beyond the Islamic world, distributed throughout Islamic lands which normally would not be able to grow these crops. Some have referred to the diffusion of numerous crops during this period as the "Globalisation of Crops", which, along with an increased mechanization of agriculture, led to major changes in economy, population distribution, vegetation cover, agricultural production and income, population levels, urban growth, the distribution of the labour force, linked industries, cooking and diet, clothing, and numerous other aspects of life in the Islamic world.

Muslim engineers in the Islamic world were responsible for numerous innovative industrial uses of hydropower, the first industrial uses of tidal power, wind power, steam power, and petroleum, and the earliest large factory complexes (*tiraz* in Arabic). The industrial uses of watermills were in widespread use since the 8th century. A variety of industrial mills were first invented in the Islamic world, including fulling mills, gristmills, hullers, paper mills, sawmills, shipmills, stamp mills, steel mills, sugar mills, tide mills, and windmills. By the 11th century, every province throughout the Islamic world had these industrial mills in operation, from al-Andalus and North Africa to the Middle East and Central Asia. Muslim engineers also invented crankshafts and water turbines, first employed gears in mills and water-raising machines, and pioneered the use of dams as a source of water power, used to provide additional power to watermills and water-raising machines. Such advances made it possible for many industrial tasks that were previously driven by manual labour in ancient times to be mechanized and driven by machinery instead in the medieval Islamic world. The transfer of these technologies to medieval Europe later laid the foundations for the Industrial Revolution in 18th century Europe.

A significant number of inventions were produced by Muslim scientists and engineers during this time, including inventors such as Abbas Ibn Firnas, Taqi al-Din, and especially al-Jazari, who is considered the "father of robotics" and "father of modern day engineering". Some of the inventions from the Islamic Golden Age include the camera obscura, coffee, hang glider, hard soap, shampoo, pure distillation, liquefaction, crystallisation, purification, oxidisation, evaporation, filtration, distilled alcohol, uric acid, nitric acid, alembic, crankshaft, valve, reciprocating suction piston pump, mechanical clocks driven by water and weights, programmable humanoid robot, combination lock, quilting, pointed arch, scalpel, bone saw, forceps, surgical catgut, windmill, inoculation, smallpox vaccine, fountain pen, cryptanalysis, frequency analysis, three-course meal, stained glass and quartz glass, Persian carpet, modern cheque, celestial globe, explosive rockets and incendiary devices, torpedo, and royal pleasure gardens.

Medieval Europe

European technology in the Middle Ages may be best described as a symbiosis of *traditio et innovatio*. While medieval technology has been long depicted as a step backwards in the evolution of Western technology, sometimes willfully so by modern authors intent on denouncing the church as antagonistic to scientific progress a generation of medievalists around the American historian of science Lynn White stressed from the 1940s onwards the innovative character of many medieval techniques. Genuine medieval contributions include for example mechanical clocks, spectacles and vertical windmills. Medieval ingenuity was also displayed in the invention of seemingly inconspicuous items like the watermark or the functional button. In navigation, the foundation to the subsequent age of exploration was laid by the introduction of pintle-and-gudgeon rudders, lateen sails, the dry compass and the astrolab.

Significant advances were also made in military technology with the development of plate armour, steel crossbows, counterweight trebuchets and cannon. Perhaps best known are the Middle Ages for their architectural heritage: While the invention of the rib vault and pointed arch gave rise to the high rising Gothic style, the ubiquitous medieval fortifications gave the era the almost proverbial title of the 'age of castles'.

Renaissance

Age of Exploration

The sailing ship (Nau or Carrack) enabled the Age of Exploration with the European colonization of the Americas, epitomized by Francis Bacon's *The New Atlantis*. European powers rediscovered the idea of the Civil code, lost since the time of the Ancient Greeks.

Industrial Revolution

The British Industrial Revolution is characterized by developments in the areas of textile manufacturing, metallurgy and transport driven by the development of the steam engine.

19th Century

The 19th century saw astonishing developments in transportation, construction, and communication technologies originiating in Europe. The Steam Engine which had existed since the early 18th century, was practically applied to both steamboat and railway transportation. Telegraphy also developed into a practical technology in the 19th century. Other technologies were explored for the first time, including the Incandescent light bulb. The Portsmouth Block Mills was where manufacture of ships' pulley blocks by all-metal machines first took place and instigated the age of mass production. Machine tools used by engineers to manufacture other machines began in the first decade of the century, notably by Richard Roberts and Joseph Whitworth. Steamships were eventually completely iron-clad, and played a role in the opening of Japan and China to trade with the West. Mechanical computing was envisioned by Charles Babbage but did not come to fruition. The Second Industrial Revolution at the end of the 19th century saw rapid development of chemical, electrical, petroleum, and steel technologies connected with highly structured technology research.

20th Century

20th Century technology developed rapidly. Communication technology, transportation technology, broad teaching and implementation of Scientific method, and increased research spending all contributed to the advancement of modern science and technology. Due to the scientific gains directly tied to military research and development, technologies including electronic computing might have developed as rapidly as they did in part due to war. Radio, Radar, and early sound recording were key technologies which paved the way for the Telephone, Fax machine, and Magnetic storage of data. Energy and engine technology improvements were also vast, including Nuclear power, developed after the Manhattan project. Transport by rocketry: most work occurred in the U.S. (Goddard), Russia (Tsiolkovsky) and Germany (Oberth). Making use of computers and advanced research labs, modern scientists have Recombinant DNA.

21st Century

Despite the fact we have just entered into the 21st century, technology is being developed even more rapidly, marked progress in almost all fields of science and technology has led to massive improvements to the technology we currently possess, the rate of development in computers being only one example at which the speed of progress continues forward, leading to the speculation of a technological singularity occurring within this century. Current ongoing developments include research into the scramjet, nanotechnology, bioengineering, nuclear fusion, new developments in armor, advanced materials and a plethora of other fields, leading to speculations among some circles of the development of devices such as powered armor in the near future.

Measuring Technological Progress

Many sociologists and anthropologists have created social theories dealing with social and cultural evolution. Some, like Lewis H. Morgan, Leslie White, and Gerhard Lenski, declare technological progress to be the primary factor driving the development of human civilization. Morgan's concept of three major stages of social evolution (savagery, barbarism, and civilization) can be divided by technological milestones, like fire, the bow, and pottery in the savage era, domestication of animals, agriculture, and metalworking in the barbarian era and the alphabet and writing in the civilization era.

Instead of specific inventions, White decided that the measure by which to judge the evolution of culture was energy. For White "the primary function of culture" is to "harness and control energy." White differentiates between five stages of human development: In the first, people use energy of their own muscles. In the second, they use energy of domesticated animals. In the third, they use the energy of plants (agricultural revolution). In the fourth, they learn to use the energy of natural resources: coal, oil, gas. In the fifth, they harness nuclear energy. White introduced a formula P=E*T, where E is a measure of energy consumed, and T is the measure of efficiency of technical factors utilizing the energy. In his own words, "culture evolves as the amount of energy harnessed per capita per year is increased, or as the efficiency of the instrumental means of putting the energy to work is increased". Russian astronomer, Nikolai Kardashev, extrapolated his theory creating the Kardashev scale, which categorizes the energy use of advanced civilizations.

Lenski takes a more modern approach and focuses on information. The more information and knowledge (especially allowing the shaping of natural environment) a given society has, the more advanced it is. He identifies four stages of human development, based on advances in the history of communication. In the first stage, information is passed by genes. In the second, when humans gain sentience, they can learn and pass information through by experience. In the third, the humans start using signs and develop logic. In the fourth, they can create symbols, develop language and writing. Advancements in the technology of communication translates into advancements in the economic system and political system, distribution of wealth, social inequality and other spheres of social life. He also differentiates societies based on their level of technology, communication and economy: 1) hunters and gatherers, 2) simple agricultural, 3) advanced agricultural, 4) industrial 5) special (like fishing societies).

Finally, from the late 1970s sociologists and anthropologists like Alvin Toffler (author of Future Shock), Daniel Bell and John Naisbitt have approached the theories of post-industrial societies, arguing that the current era of industrial society is coming to an end, and services and information are becoming more important than industry and goods. Some of the more extreme visions of the post-industrial society, especially in fiction, are strikingly similar to the visions of near and post-Singularity societies.

By Type of Technology

History of Biotechnology

To be incorporated into main article:

- Timeline of agriculture and food technology
- Hunter-gatherer
- Agriculture
- Food science
- Genetically modified food
- History of agricultural science
- History of gardening
- Biotechnology (timeline, etc.)
- History of sushi
- History of tea in China

History of Civil Engineering

To be incorporated:

- Civil engineering
- Architecture and building construction
- Bridges, harbors, tunnels, dams
- Surveying, instruments and maps, cartography, urban engineering, water supply and sewerage

History of Communication

To be incorporated:

- Communications
- Writing systems
- Telecommunications
- History of mobile phones
- History of animation
- History of broadcasting
- History of radar

- History of radio
- Printing
- Cinema
- Radio
- Television
- Internet

History of Consumer Technology

To be incorporated:

- Timeline of lighting technology
- History of textiles and clothing
- History of materials science
- Family and consumer science
- History of knitting
- History of lensmaking
- History of the chair
- History of the umbrella
- Manufacturing

History of Electrical Engineering

To be incorporated:

- History of street lighting in the United States

History of Energy Technology

To be incorporated:

- Energy (History, Use by humans, See also)
- History of coal mining
- History of perpetual motion machines
- Timeline of steam power
- Timeline of alcohol fuel
- Timeline of nuclear fusion

History of Materials Science

To be incorporated:

- Timeline of materials technology
- Metallurgy
- Materials and processing

History of Measurement

To be incorporated:

- History of time in the United States
- Timeline of time measurement technology

History of Military Technology

To be incorporated into main article:

- Military history#Technological Evolution
- Category:Military history - articles on history of specific technologies

History of Nuclear Technology

- Manhattan Project
- Atomic Age
- Nuclear testing
- Nuclear arms race

History of Science and Technology

- History of telescopes
- Timeline of telescopes, observatories, and observing technology
- Timeline of microscope technology
- Timeline of particle physics technology
- Timeline of low-temperature technology
- Timeline of temperature and pressure measurement technology

History of Transport Technology

To be incorporated into main article:

- Timeline of motor and engine technology

- Timeline of photography technology
- Timeline of rocket and missile technology
- Timeline of communication technology

BIO SCIENCE TECHNOLOGY INTRODUCTION

PREREQUISITES: Degree in any biological science (including biochemistry and applied sciences such as medicine, pharmacology, etc.) or in chemical or physical sciences with a strong interest in biology.

ASSESSMENT: By critical report and written exam. A common topic will be chosen from a list of various aspects of genomic technologies. The students will be asked to submit a critical report on e.g., the state of technology, its application and limitations. Reports will be double marked by two of the module tutors. Marks will be given for literature review, content, presentation, clarity and will account for 3% of the final degree marks. The part of the closed exam allocated to genomics will make up the final part of the module assessment. The closed exam will be anonymously marked and account for 3% of the final degree marks. The closed exam will last 1 hour and consist of a mixture of short-answer, factual questions and a longer-answer, problem solving section.

SUMMARY: The module will teach and train students about various aspects of genomic technologies. Over 2 weeks, students will obtain both a theoretical appreciation of the technologies as well as gaining hands-on experience of gene expression and genotyping systems. The students will work in a step-by-step manner through gene expression systems, post analysis, real-time PCR and various genotyping methodologies, concentrating on machine optimisation and maximising machine output. Each aspect will be covered in detail and with real samples used in the practical tutorials.

AIMS: The module aim will be to provide each student with an in-depth knowledge of the state-of the-art genomic technologies. They will gain hands-on experience of the machines used for genome analysis and learn advanced practical applications associated with these technologies. Students will also be trained in various data analysis techniques related to expression analysis and genotyping.

LEARNING OUTCOMES: Students will know and understand the theory and principles behind each technology. They will be able to set up instruments for various applications and understand the limitations of each technology. They will have an appreciation of the advantages and disadvantages of each technology and applications. Students will have an awareness of other emerging technologies and their integration with the current systems. They will also be equipped with in-depth knowledge of the data analysis basics and analysis issues relating to genomic technologies.

SYNOPSIS:

Week 1

Lecture 1 - Introduction to Genomics (NA)

"Genomics" – *what does it mean?* Introduction to expression array technology and genotyping– types and advantages. (NA)

Lecture 2 – Transcriptomics – theory and applications (HVI)

What is a transcriptome? Uses and applications.

Lecture 3 – Experimental design considerations (NA)

Fundamentals of experimental design in gene expression experiments. 1.5 hours

Tour of the Genomcis Lab and reading on technologies

Directed reading on specific subject technologies and various aspects

Lecture 4 – Affymetrix array system (NA)

Study the details of the Affymetrix gene chip technology, its multi-probe strategy and basics of chip fabrication.

Practical 1a+d – Affymetrix system hardware

Run as 2 groups, 45 minutes each – introduction to the Affymetrix hardware and familiarising the students with different aspects of the related equipment.

Lecture 5 – Affymetrix data exploration (NA)

Fundamentals of Affymetrix data analysis will be discussed.

Lecture 6 – Custom / spotted arrays – uses and fabrication (NA)

Fundamentals of array fabrication including available array spotting technologies, glass slide surface chemistries and substrates. Also, issues to consider will generating spotted arrays – Temperature, humidity, spotting buffers, machine controls etc.

Practical 2a+b – RNA QC and cDNA generation

Run as 2 groups, 2 hours each – focusing on RNA quality control and generating cDNA from sample RNA.

Practical 3a+b – QArray Mini

Run as 2 groups - one hour per group. Practical demonstration of the QArray Mini spotting system - machine set-up, printing, pre and post printing methodologies.

Lecture 7 – Microarray hybridisation and image processing (NA)

Fundamentals of the microarray hybridisation process, detection of spots and image processing.

Practical 4a+b – Generated cDNA clean-up and labelling setup

Run as 2 groups, 1.5 hours each – continue with clean-up processes for the cDNA generated in Practical 3 and - target labelling of the cDNA generated

Practical 5a+b – Labelled cDNA clean-up and hybridisation

Run as 2 groups, 2 hours each - continue with clean-up processes for the labelled cDNA generated in Practical 5, including cDNA clean-up and hybridising the target to probes bound to the microarray chip.

Practical 6a-d – Microarray image scanning

Run as 4 groups, 1 hour each – computer/scanner based activity to scan the chips produced in Practical 6.

Practical 7 – Array image processing essentials

Run as a single groups - 2 hours session to learn about image acquisition and analysis. Requires B102A

Week 2

Lecture 8 – Microarray applications (IAG)

Case study 1.

Lecture 9 - Microarray applications (HVI)

Case study 2.

Practical 8 – Computer based microarray analysis

Run as 1 groups, 3 hours – computer based session to analyse and transform the data produced in Practical 8. Requires B102A or A004

Lecture 10 - Real-Time PCR basics and experimental essentials (NA)

Introduction to the Real-Time (quantitative) PCR system, focusing on design considerations and various available chemistries. Introduction to experimental set-up used in Practical 10, including experimental design, samples used and analysis system. 1.5 hours

Practical 9a+b – Real-Time plate set-up

Run as 2 groups, 1 hour each – hands-on sample / plate set for the Real-Time PCR experiment explained in Lecture 9.

Practical 10 – Real-Time PCR data retrieval and analysis

Run as 1 group, 1.5 hours – computer based, focusing on how to retrieve your data once the QPCR is run and how to analyse your results, including normalisation and calculation of fold-changes. Requires B102A.

Lecture 11 – Introduction to sequencing (PDA)

Fundamentals of sequencing.

Lecture 12 – SNP detection and Genotyping (NA)

Fundamentals of genotyping with particular focus on SNP detection.

Bioinformatics Lecture L3 – Assigning function to sequence (PDA)

Fundamentals of assigning function to DNA sequences

Practical 11a+b - ABI 3130 demonstration

Run as 2 groups - 45 minutes each. Practical demonstration of the ABI 3130 Genetic analyser - machine components, set-up and sequencing chemistries.

Bioinformatics Workshop W5 – Sequencing analysis (PDA)

Detailed tutorial (3 hours) on analysing the sequencing results and trouble shooting potential problems. Requires B102A

Reading/ Surgery

Assessment by critical report – Submission date: 16.30 Monday 19 November 2007

RECOMMENDED READING

Key Text:

- DNA Microarrays (2006). Mark Schena (Editor), Scion Publishing Ltd,

Other Useful Texts:

- Microarray Biochip Technology (2000). M. Schena (Editor), Eaton Publishing Company/Biotechniques Books,
- A biologists guide to Analysis of DNA Microarray Data (2002) by Steen Knudsen, John Wiley & Sons INC,
- Microarray Bioinformatics (2003) by Dov Stekel, Cambridge University Press,

LECTURERS AND ORGANISATION

Lectures

NA Dr Naveed Aziz

HVI Dr Harv Isaacs

IAG Professor Ian Graham

PDA Dr Peter Ashton

Practicals

Practicals 1-11 all given by NA and CW (Celina Whalley) using the Genomics labs or Bioinformatics Suite B102A

Maximum Numbers: 16

Student Load

Lectures: 13 hours

Practicals: 17 hours

Total Contact Hours : 30

Exam : 5.5

Private Study: 64.5

STAFF TEACHING COMMITMENTS: *Please enter the total number of sessions attended by each staff member:*

Staff initials	*NA*	*CW*	*HVI*	*IAG*	*PDA*
Lecture sess.	8		2	1	1
Practical sess.	11	11			
Other (specify)					

SAFETY AND TIMETABLING INFORMATION: *Please fill in an entry in the table below for each teaching session in the module. Where possible group sessions that have identical entries in all columns. A key below explains the column headings.*

Sess.	*Type*	*Occ.*	*Haz.*	*Max.*	*Equipment*	*Lecturers*
Lectures 1,2,4-9, 11-12	L	1	A	16	Powerpoint presentation	NA, HVI, IAG, PA
Lecture 3 + 10	L	1	A	16	Powerpoint presentation	NA

Practicals 7, 8, 10	P	1	B	16	Computers, requires B102A	NA, CW
Practicals 1, 2, 3, 4,	P	2	C	8	TF labs	NA, CW
Practicals 6	P	4	C	4	TF Labs and Equipment	NA, CW

KEY: Sess: session number or group of sessions (e.g. 6, 8-13). Type: Type & Duration; L 1 hr-lecture, P 3 hr-practical class, S 1 hr-seminar, T 1 hr tutorial. Specify non-standard type or duration. Occ.: number of occurrences of the session (e.g. 3 occurrences, each taking one third of the class). Haz.: hazard rating: A, low hazard, lectures, 'paper & pencil' problem sessions, etc; B, medium hazard, observational practicals where students move about but are not involved in C category activities; C, high hazard, practicals involving potential hazard in overcrowded laboratories, e.g. naked flames, hot liquids, glassware, pipetting. Max.: maximum number of students permitted in session, which may be less than the maximum number taking the module in, for example, a circus practical (see maximum room capacities above). Equipment: essential equipment in limited supply: M microscopes, D dissecting microscopes, C computers, S spectrophotometers, Ch chart recorders, H haemocytometers, Cm microcentrifuges, Cc cooled centrifuges, Cb bench centrifuges, G Gilson pipettes, specify other items. Lecturers: the initials of the lecturers participating in the session. Description: a brief description of the session(s).

REFERENCES

Adam Robert Lucas (2005), "Industrial Milling in the Ancient and Medieval Worlds: A Survey of the Evidence for an Industrial Revolution in Medieval Europe", *Technology and Culture* 46 (1), p. 1-30.

Adam Robert Lucas (2005), "Industrial Milling in the Ancient and Medieval Worlds: A Survey of the Evidence for an Industrial Revolution in Medieval Europe", *Technology and Culture* 46 (1), p. 1-30.

Ahmad Y Hassan, Transfer Of Islamic Technology To The West, Part II: Transmission Of Islamic Engineering.

Andrew M. Watson (1974), "The Arab Agricultural Revolution and Its Diffusion, 700-1100", *The Journal of Economic History* 34 (1), p. 8-35.

Andrew M. Watson (1983), *Agricultural Innovation in the Early Islamic World*, Cambridge University Press.

Brush, S. G. (1988). *The History of Modern Science: A Guide to the Second Scientific Revolution 1800-1950*. Ames: Iowa State University Press.

Bunch, Bryan and Hellemans, Alexander, (1993) *The Timetables of Technology,* New York, Simon and Schuster.

Derry, Thomas Kingston and Williams, Trevor I., (1993) *A Short History of Technology: From the Earliest Times to A.D. 1900*. New York: Dover Publications.

Greenwood, Jeremy (1997) *The Third Industrial Revolution: Technology, Productivity and Income Inequality* AEI Press.

Kranzberg, Melvin and Pursell, Carroll W. Jr., eds. (1967)*Technology in Western Civilization: Technology in the Twentieth Century* New York: Oxford University Press.

Olby, R. C. et al., eds. (1996). *Companion to the History of Modern Science,*. New York, Routledge.

Pacey, Arnold, (1974, 2ed 1994),*The Maze of Ingenuity* The MIT Press, Cambridge, Mass, 1974, [2ed 1994, cited here].

Paul Vallely, How Islamic Inventors Changed the World, *The Independent*, 11 March 2006.

Singer, C., Holmyard, E.J., Hall, A. R and Williams, T. I. (eds.), (1954-59 and 1978) *A History of Technology,*, 7 vols., Oxford, Clarendon Press,. (Vols 6 and 7, 1978, ed. T. I. Williams).

Thomas F. Glick (1977), “Noria Pots in Spain”, *Technology and Culture* 18 (4), p. 644-650.

7

Overview of Biological Technologies with Select Case Studies

BIOLOGICAL TECHNOLOGY

Biological technology is technology based on biology, especially when used in agriculture, food science, and medicine. The United Nations Convention on Biological Diversity has come up with one of many definitions of biotechnology: Biotechnology has the potential to assist farmers in reducing on-farm chemical inputs and produce value-added commodities. Conversely, there are concerns about the use of biotechnology in agricultural systems including the possibility that it may lead to greater farmer dependence on the providers of the new technology. Different types of crops have been produced using the molecular tools of biotechnology and are beginning to be utilized in agricultural systems all over the world.

At the same time, an increasing number of farmers are adopting sustainable cultural practices. Early man was a food gatherer, depending on nature for all his needs. He gradually moved on to being a food grower with the discovery of agriculture, and settled down in one place, learning to live in a group. This was the beginning of civilisation as we know it today. Early knowledge of agriculture was an accumulation of experiences that were passed on from father to son. Some of these have been preserved as religious commandments and some in the ancient inscriptions. There is evidence to show that as early as 2000 BC the Egyptian civilisation followed particular dates for sowing and reaping. Some Greek and Roman classics give instructions on how to get a higher yield.

The development of agriculture made it apparent that more food could be extracted from a given area of land by encouraging useful and hardy plant and animal species, and discouraging others.

At the turn of the 19th century, a movement began in central Europe to train farmers in specific farming skills. A truly scientific approach was begun by Justine von Liebig of Darmstadt who in his classic work introduced the systematic development of agriculture science. From the 19th century onwards plant production became a scientific discipline.

In the early 20th century, the legendary work of Gregor Mendel laid the foundation of modern day genetics. His work explained the basics of inheritance in terms of the factor we today call genes.

Apart from selection and hybridisation, new and innovative techniques such as genetic engineering that aid plant breeders have been developed in the recent past. One example of this is BtCotton. With advances in human and plant biology, more intricate details about the cell – the basic unit of life – were illuminated. The possibility of raising whole plants from various plant tissues, commonly know as tissue culture, has thrown open the doors for expedited evolution both in terms of generation of genetic variability and multiplication of elite plant types. The knowledge of the wonder molecule DNA has also opened a new area of plant breeding research. These new technologies have been collectively referred to as biotechnology. It is a collective effort for plant breeding in the future and will compliment man's crusade for more and better food. In India, the Green Revolution saw the rapid progress of agriculture and the application of different methods to enhance production. Biofertilizers have been proven to be more environmentally friendly fertilizers that do not cause harm to life. Bioremediation methods have been used to clear oil spills using bacteria. Biotechnology means any technological application that uses biological systems, living organisms, or derivatives thereof, to make or modify products or processes for specific use. Biological technology is technology based on biology, especially when used in agriculture, food science, and medicine.

In fact, the term should be used in a much broader sense to describe the whole range of methods, both ancient and modern, used to manipulate organic matter to meet human needs. So the term can be defined as, "The application of indigenous and/or scientific knowledge to the management of (parts of) microorganisms, or of cells and tissues of higher organisms, so that these supply goods and services of use to human beings. Before the 1970s, the term, *biotechnology*, was primarily used in the food processing and agriculture industries. Since the 1970s, it began to be used by the Western scientific establishment to refer to laboratory-based techniques being developed in biological research, such as recombinant DNA or tissue culture-based processes. There has been a great deal of talk - and money - poured into biotechnology with the hope that miracle drugs will appear. While there do seem to be a small number of efficacious drugs, in general the biotech revolution has not happened in the pharmaceutical sector. Biotechnology combines disciplines like genetics, molecular biology, biochemistry, embryology and cell biology, which are in turn linked to practical disciplines like chemical engineering, information technology, and robotics. However, recent progress with monoclonal antibody based drugs, such as Genentech's Avastin (tm) suggest that biotech may finally have found a role in pharmaceutical sales. Biotechnology can also be defined as the manipulation of organisms to do practical things and to provide useful products. Early cultures also understood the importance of using natural processes to breakdown waste products into inert forms. From very early nomadic tribes to pre-urban civilisations it was common knowledge that given enough time organic waste products would be absorbed and eventually integrated into the soil.

It was not until the advent of modern microbiology and chemistry that this process was fully understood and attributed to bacteria. The most practical use of biotechnology, which is still present today, is the cultivations of plants to produce food suitable to humans. Agriculture has been theorized to have become the dominant way of producing food since the Neolithic Revolution. The processes and methods of agriculture have been refined by other mechanical and biological sciences since its inception. Through early biotechnology farmers were able to select the best suited and high-yield crops to produce enough food to support a growing population.

Other uses of biotechnology were required as crops and fields became increasingly large and difficult to maintain. Specific organisms and organism byproducts were used to fertilize, restore nitrogen, and control pests. Throughout the use of agriculture farmers have inadvertently altered the genetics of their crops through introducing them to new environments, breeding them with other plants, and by using artificial selection. In modern times some plants are genetically modified to produce specific nutritional values or to be economical.

The process of Ethanol fermentation was also one of the first forms of biotechnology. Cultures such as those in Mesopotamia, Egypt, and Iran developed the process of brewing which consisted of combining malted grains with specific yeasts to produce alcoholic beverages. In this process the carbohydrates in the grains were broken down into alcohols such as ethanol. Later other cultures produced the process of Lactic acid fermentation which allowed the fermentation and preservation of other forms of food. Fermentation was also used in this time period to produce leavened bread. Although the process of fermentation was not fully understood until Louis Pasteur's work in 1857, it is still the first use of biotechnology to convert a food source into another form.

Combinations of plants and other organisms were used as medications in many early civilisations. Since as early as 200 BC people began to use disabled or minute amounts of infectious agents to immunize themselves against infections. These and similar processes have been refined in modern medicine and have lead to many developments such as antibiotics, vaccines, and other methods of fighting sickness.

In the early twenthieth century, our society gained a greater understanding of biochemical and genetic mechanisms, and we began to explore ways of manufacturing specific products using microbiology techniques. In 1917, Chaim Weizmann first used a pure culture in an industrial process, that of manufacturing corn starch using *Clostridium acetobutylicum* to produce acetone, which the United Kingdom desperately needed to manufacture explosives during World War I.

The field of modern biotechnology is thought to have largely began on June 16, 1980, when the United States Supreme Court ruled that a genetically-modified microorganism could be patented in the case of *Diamond v. Chakrabarty*. Indian-born Ananda Chakrabarty, working for General Electric, had developed a bacterium (derived from the *Pseudomonas* genus) capable of breaking down crude oil, which he proposed to use in treating oil spills.

There are four important technical terms to know when dealing with biotechnology: genetics, genes, genome and genetically modified organisms. Genetics is the branch of biology that deals with the principles of heredity and variation in all living things. It is the study of why and how parents pass on some of their distinguishing features to their offspring. Its focus is on genes and their functions. The gene is the basic unit of heredity and the ultimate arbiter of what we are. It carries instructions that allow cells to produce specific proteins. (It should be noted, however, that only certain genes are active at any given moment and environment.) A gene is a part of the deoxyribonucleic acid (DNA) molecule. DNA, which is present in all living cells, contains information coding for cellular structure, organisation and function. It is made up of two strands twisted around each other in a helical staircase.

Each cell in an organism has one or two sets of the basic DNA complement, called a genome. The genome is itself made up of one or more extremely long linear array of molecules of DNA that are called chromosomes. Genes, as explained earlier, are the functional regions of the DNA. They are the active segments of the chromosomes. In modern biotechnology, the genome of an organism is altered by exposing cells to fragments of "foreign" DNA carrying the desirable genes, often from another species.

This DNA is taken in and inserts itself into one or more of the recipient's chromosomes at a location where it is inherited like any other part of the genome. The cells so modified are called transgenic cells. It is from transgenic cells that a GMO can be produced. All of the GMO's cells contain the additional foreign DNA.

There is no universal definition for genetically modified organism (also called "transgenic organism" or "living modified organism"). However, it is generally understood to be a plant, animal or microorganism that contains genes that have been altered or transferred from another species or from the same species by means of genetic engineering techniques.

The role of information technology in the development of modern biotechnology Knowledge of biology has rapidly grown over the years, requiring the development of powerful tools to handle all of it. Information technology, through the field of bioinformatics, makes possible the rapid organisation and analysis of biological data. Bioinformatics merges biology, computer science, and information technology to manage and analyze genomic data, with the ultimate goal of understanding and modeling living systems.

There are many potential benefits that modern biotechnology offers humankind in general. The European Commission (2002) refers to modern biotechnology as the "next wave of the knowledge-based economy" after information technology, and the "most promising of the frontier technologies." In its broadest sense, "biotechnology" refers to "any technique that uses living organisms, or parts of such organisms, to make or modify products, to improve plants or animals, or to develop microorganisms for specific use." Biotechnology has evolved through the years.

On one end of the development pole are techniques of traditional biotechnology like microbial fermentation, used as early as 10,000 years ago in fermenting beer, wine and dairy products. At the other end of the development pole are the continuously evolving techniques of modern biotechnology, such as genetic engineering. Using genetic engineering techniques, the genetic makeup of an organism may be modified by inactivating or altering some of its genes and introducing other natural or artificial genes, usually from another organism.

The Cartagena Protocol on Biosafety defines modern biotechnology as referring to any process that involves the application of (i) *in vitro* nucleic acid techniques, including recombinant deoxyribonucleic acid and direct injection of nucleic acid into cells or organelles, or (ii) fusion of cells beyond the taxonomic family, that overcome natural physiological reproductive or recombination of barriers and that are not techniques used in traditional breeding and selection. Although the Protocol is not yet in force (because less than the required 50 States have either ratified or acceded to it), the Protocol's definition of modern biotechnology has gained currency in international circles.

However, while there may be an emerging international consensus on the above definition, strictly speaking it is a definition that is applicable only when one uses the term "modern biotechnology" for purposes of interpreting or implementing the Protocol.

One application of biotechnology is the directed use of organisms for the manufacture of organic products (examples include beer and milk products). Another example is using naturally present bacteria by the mining industry in bioleaching. Biotechnology is also used to recycle, treat waste, clean up sites contaminated by industrial activities (bioremediation), and produce biological weapons. Red biotechnology is applied to medical processes. Some examples are the designing of organisms to produce antibiotics, and the engineering of genetic cures through genomic manipulation. White biotechnology also known as grey biotechnology, is biotechnology applied to industrial processes. An example is the designing of an organism to produce a useful chemical.

Another example is the using of enzymes as industrial catalysts to either produce valuable chemicals or destroy harzadous/polluting chemicals (examples using oxidoreducatses are given in Feng Xu (2005) "Applications of oxidoreductases: Recent progress" Ind. Biotechnol. 1, 38-50). White biotechnology tends to consume less in resources than traditional processes used to produce industrial goods. Green biotechnology is biotechnology applied to agricultural processes. An example is the designing of transgenic plants to grow under specific environmental conditions or in the presence (or absence) of certain agricultural chemicals.

One hope is that green biotechnology might produce more environmentally friendly solutions than traditional industrial agriculture. An example of this is the engineering of a plant to express a pesticide, thereby eliminating the need for external application of pesticides. An example of this would be Bt corn. Whether or not green biotechnology products such as this are ultimately more environmentally friendly is a topic of considerable debate. Bioinformatics is an interdisciplinary field which addresses biological problems using computational techniques. The field is also often referred to as computational biology. It plays a key role in various areas, such as functional genomics, structural genomics, and proteomics, and forms a key component in the biotechnology and pharmaceutical sector.

The term blue biotechnology has also been used to describe the marine and aquatic applications of biotechnology, but its use is relatively rare. One is improved yield from crops. Using the techniques of modern biotechnology, one or two genes may be transferred to a highly developed crop variety to impart a new character that would increase its yield. However, while increase in crop yield is the most obvious application of modern biotechnology in agriculture, it is also the most difficult one. Current genetic engineering techniques work best for effects that are controlled by a single gene. Many of the genetic characteristics associated with yield (*e.g.*, enhanced growth) are controlled by a large number of genes, each of which has a minimal effect on the overall yield. There is, therefore, much scientific work to be done in this area.

Another is the reduced vulnerability of crops to environmental stresses. Crops containing genes that will enable them to withstand biotic and abiotic stresses may be developed. For example, drought and excessively salty soil are the two most important limiting factors in crop productivity. Biotechnologists are studying plants that can cope with these extreme conditions in the hope of finding the genes that enable them to do so and eventually transferring these genes to the more desirable crops. One of the latest developments is the identification of a plant gene, At-DBF2, from thale cress, a tiny weed that is often used for plant research because it is very easy to grow and its genetic code is well mapped out. When this gene was inserted into tomato and tobacco cells, the cells were able to withstand environmental stresses like salt, drought, cold and heat, far more than ordinary cells. If these preliminary results prove successful in larger trials, then At-DBF2 genes can help in engineering crops that can better withstand harsh environments.

Researchers have also created transgenic rice plants that are resistant to rice yellow mottle virus (RYMV). In Africa, this virus destroys majority of the rice crops and makes the surviving plants more susceptible to fungal infections.

Improved taste, texture or appearance of food. Modern biotechnology can be used to slow down the process of spoilage so that fruit can ripen longer on the plant and then be transported to the consumer with a still reasonable shelf life. This improves the taste, texture and appearance of the fruit. More importantly, it could expand the market for farmers in developing countries due to the reduction in spoilage.

The first genetically modified food product was a tomato which was transformed to delay its ripening. Researchers in Indonesia, Malaysia, Thailand, Philippines and Vietnam are currently working on delayed-ripening papaya in collaboration with the University of Nottingham and Zeneca.

Reduced dependence on fertilizers, pesticides and other agrochemicals. Most of the current commercial applications of modern biotechnology in agriculture are on reducing the dependence of farmers on agrochemicals. For example, Bacillus thuringiensis (Bt) is a soil bacterium that produces a protein with insecticidal qualities. Traditionally, a fermentation process has been used to produce an insecticidal spray from these bacteria. In this form, the Bt toxin occurs as an inactive protoxin, which requires digestion by an insect to be effective. There are several Bt toxins and each one is specific to certain target insects. Crop plants have now been engineered to contain and express the genes for Bt toxin, which they produce in its active form. When a susceptible insect ingests the transgenic crop cultivar expressing the Bt protein, it stops feeding and soon thereafter dies as a result of the Bt toxin binding to its gut wall. Bt corn is now commercially available in a number of countries to control corn borer, which is otherwise controlled by spraying.

Crops have also been genetically engineered to acquire tolerance to broad-spectrum herbicide. The lack of cost-effective herbicides with broad-spectrum activity and no crop injury was a consistent limitation in crop weed management. Multiple applications of numerous herbicides were routinely used to control a wide range of weed species detrimental to agronomic crops. Weed management tended to rely on preemergence — that is, herbicide applications were sprayed in response to expected weed infestations rather than in response to actual weeds present. Mechanical cultivation and hand weeding were often necessary to control weeds not controlled by herbicide applications. The introduction of herbicide tolerant crops has the potential of reducing the number of herbicide active ingredients used for weed management, reducing the number of herbicide applications made during a season, and increasing yield due to improved weed management and less crop injury. Transgenic crops that express tolerance to glyphosphate, glufosinate and bromoxynil have been developed. These herbicides can now be sprayed on transgenic crops without inflicting damage on the crops while killing nearby weeds.

From 1996 to 2001, herbicide tolerance was the most dominant trait introduced to commercially available transgenic crops, followed by insect resistance. In 2001, herbicide tolerance deployed in soybean, corn and cotton accounted for 77% of the 62.6 million hectares planted to transgenic crops; Bt crops accounted for 15%; and stacked genes for herbicide tolerance and insect resistance used in both cotton and corn accounted for 8%.

Production of novel substances in crop plants. Modern biotechnology is increasingly being applied for novel uses other than food. For example, oilseed is at present used mainly for margarine and other food oils, but it can be modified to produce fatty acids for detergents, substitute fuels and petrochemicals. Banana trees and tomato plants have also been genetically engineered to produce vaccines in their fruit. If future clinical trials prove successful, the advantages of edible vaccines would be enormous, especially for developing countries. The transgenic plants may be grown locally and cheaply. Homegrown vaccines would also avoid logistical and economic problems posed by having to transport traditional preparations over long distances and keeping them cold while in transit. And since they are edible, they will not need syringes, which are not only an additional expense in the traditional vaccine preparations but also a source of infections if contaminated.

In medicine, modern biotechnology finds promising applications in: (i) pharmacogenomics; (ii) drug production; (iii) genetic testing; and (v) gene therapy. Pharmacogenomics is the study of how the genetic inheritance of an individual affects his/her body's response to drugs. It is a coined word derived from the words "pharmacology" and "genomics". It is therefore the study of the relationship between pharmaceuticals and genetics. The vision of pharmacogenomics is to be able to design and produce drugs that are adapted to each person's genetic makeup.

Pharmacogenomics results in the following benefits:

(i) Development of tailor-made medicines. Using pharmacogenomics, pharmaceutical companies can create drugs based on the proteins, enzymes and RNA molecules that are associated with specific genes and diseases. These tailor-made drugs promise not only to maximize therapeutic effects but also to decrease damage to nearby healthy cells.

(ii) More accurate methods of determining appropriate drug dosages. Knowing a patient's genetics will enable doctors to determine how well his/ her body can process and metabolize a medicine. This will maximize the value of the medicine and decrease the likelihood of overdose.

(iii) Improvements in the drug discovery and approval process. The discovery of potential therapies will be made easier using genome targets. Genes have been associated with numerous diseases and disorders. With modern biotechnology, these genes can be used as targets for the development of effective new therapies, which could significantly shorten the drug discovery process.

(iv) Better vaccines. Safer vaccines can be designed and produced by organisms transformed by means of genetic engineering. These vaccines will elicit the immune response without the attendant risks of infection. They will be inexpensive, stable, easy to store, and capable of being engineered to carry several strains of pathogen at once.

Traditional pharmaceutical drugs are small chemicals molecules that treat the symptoms of a disease or illness - one molecule directed at a single target. Biopharmaceuticals are large biological molecules known as proteins and these target the underlying mechanisms and pathways of a malady; it is a relatively young industry. They can deal with targets in humans that are not accessible with traditional medicines. A patient typically is dosed with a small molecule *via* a tablet while a large molecule is typically injected. Small molecules are manufactured by chemistry but large molecules are created by living cells: for example — bacteria cells, yeast cell,animal cells. Modern biotechnology is often associated with the use of genetically altered microorganisms such as *E. coli* or yeast for the production of substances like insulin or antibiotics.

It can also refer to transgenic animals or transgenic plants, such as Bt corn. Genetically altered mammalian cells, such as Chinese Hamster Ovary (CHO) cells, are also widely used to manufacture pharmaceuticals. Another promising new biotechnology application is the development of plant-made pharmaceuticals. Biotechnology is also commonly associated with landmark breakthroughs in new medical therapies to treat diabetes, hepatitis B, hepatitis C, cancers, arthritis, haemophilia, bone fractures, multiple sclerosis, cardiovascular as well as molecular diagnostic devices than can be used to define the patient population. Herceptin, is the first drug approved for use with a matching diagnostic test and is used to treat breast cancer in women whose cancer cells express the protein HER2. Modern biotechnology can be used to manufacture existing drugs more easily and cheaply. The first genetically engineered products were medicines designed to combat human diseases.

To cite one example, in 1978 Genentech joined a gene for insulin and a plasmid vector and put the resulting gene into a bacterium called Escherichia coli. Insulin, widely used for the treatment of diabetes, was previously extracted from sheep and pigs. It was very expensive and often elicited unwanted allergic responses. The resulting genetically engineered bacterium enabled the production of vast quantities of human insulin at low cost. Since then modern biotechnology has made it possible to produce more easily and cheaply the human growth hormone, clotting factors for hemophiliacs, fertility drugs, erythropoietin and other drugs.

Most drugs today are based on about 500 molecular targets. Genomic knowledge of the genes involved in diseases, disease pathways, and drug-response sites are expected to lead to the discovery of thousands more new targets.

Biotechnological engineering or biological engineering is a branch of engineering that focuses on biotechnologies and biological science. It includes different disciplines such as biochemical engineering, biomedical engineering, bio-process engineering, biosystem engineering and so on. Because of the novelty field, the definition of a bioengineer is still ill defined. However, in general it is an integrated approach of fundamental biological sciences and traditional engineering principles.

Bioengineers are often employed to scale up a bio processes from the laboratory scale to the manufacturing scale. Moreover, as with most engineers, they often deal with management, economic and legal issues. Since patents and regulation (*e.g.* FDA regulation in the U.S.) are very important issues for biotech enterprises, bioengineers are often required to have knowledge related to these issues.

The increasing number of biotech enterprises is likely to create a need for bioengineers in the years to come. Many universities throughout the world are now providing programs in bioengineering and biotechnology (as indepedent programs or specialty programs within more established engineering fields).

Genetic testing involves the direct examination of the DNA molecule itself. A scientist scans a patient's DNA sample for mutated sequences.

There are two major types of gene tests. In the first type, a researcher may design short pieces of DNA ("probes") whose sequences are complementary to the mutated sequences. These probes will seek their complement among the base pairs of an individual's genome. If the mutated sequence is present in the patient's genome, the probe will bind to it and flag the mutation. In the second type, a researcher may conduct the gene test by comparing the sequence of DNA bases in a patient's gene to a normal version of the gene.

Some genetic tests are already available, although most of them are used in developed countries. The tests currently available can detect mutations associated with rare genetic disorders like cystic fibrosis, sickle cell anemia, and Huntington's disease. Recently, tests have been developed to detect mutation for a handful of more complex conditions such as breast, ovarian, and colon cancers. However, gene tests may not detect every mutation associated with a particular condition because many are as yet undiscovered, and the ones they do detect may present different risks to different people and populations.

Gene therapy may be used for treating, or even curing, genetic and acquired diseases like cancer and AIDS by using normal genes to supplement or replace defective genes or to bolster a normal function such as immunity. It can be used to target somatic (*i.e.*, body) or germ (*i.e.*, egg and sperm) cells. In somatic gene therapy, the genome of the recipient is changed, but this change is not passed along to the next generation. In contrast, in germline gene therapy, the egg and sperm cells of the parents are changed for the purpose of passing on the changes to their offspring.

There are basically two ways of implementing a gene therapy treatment:

1. *Ex vivo*, which means "outside the body" – Cells from the patient's blood or bone marrow are removed and grown in the laboratory. They are then exposed to the virus carrying the desired gene. The virus enters the cells, and the desired gene becomes part of the DNA of the cells. The cells are allowed to grow in the laboratory before being returned to the patient by injection into a vein.

2. *In vivo*, which means "inside the body" – No cells are removed from the patient's body. Instead, vectors are used to deliver the desired gene to cells in the patient's body.

Currently, the use of gene therapy is limited. Somatic gene therapy is primarily at the experimental stage. Germline therapy is the subject of much discussion but it is not being actively investigated in larger animals and human beings.

As of June 2001, more than 500 clinical gene-therapy trials involving about 3,500 patients have been identified worldwide. Around 78% of these are in the United States, with Europe having 18%. These trials focus on various types of cancer, although other multigenic diseases are being studied as well. Recently, two children born with severe combined immunodeficiency disorder ("SCID") were reported to have been cured after being given genetically engineered cells.

Gene therapy faces many obstacles before it can become a practical approach for treating disease. At least four of these obstacles are as follows:

1. *Gene delivery tools*. Genes are inserted into the body using gene carriers called vectors. The most common vectors now are viruses, which have evolved a way of encapsulating and delivering their genes to human cells in a pathogenic manner. Scientists manipulate the genome of the virus by removing the disease-causing genes and inserting the therapeutic genes. However, while viruses are effective, they can introduce problems like toxicity, immune and inflammatory responses, and gene control and targeting issues.
2. *Limited knowledge of the functions of genes*. Scientists currently know the functions of only a few genes. Hence, gene therapy can address only some genes that cause a particular disease. Worse, it is not known exactly whether genes have more than one function, which creates uncertainty as to whether replacing such genes is indeed desirable.
3. *Multigene disorders and effect of environment*. Most genetic disorders involve more than one gene. Moreover, most diseases involve the interaction of several genes and the environment. For example, many people with cancer not only inherit the disease gene for the disorder, but may have also failed to inherit specific tumor suppressor genes. Diet, exercise, smoking and other environmental factors may have also contributed to their disease.
4. *High costs*. Since gene therapy is relatively new and at an experimental stage, it is an expensive treatment to undertake. This explains why current studies are focused on illnesses commonly found in developed countries, where more people can afford to pay for treatment. It may take decades before developing countries can take advantage of this technology.

The Human Genome Project is an initiative of the U.S. Department of Energy ("DOE") that aims to generate a high-quality reference sequence for the entire human genome and identify all the human genes.

The DOE and its predecessor agencies were assigned by the U.S. Congress to develop new energy resources and technologies and to pursue a deeper understanding of potential health and environmental risks posed by their production and use. In 1986, the DOE announced its Human Genome Initiative. Shortly thereafter, the DOE and National Institutes of Health developed a plan for a joint Human Genome Project ("HGP"), which officially began in 1990.

The HGP was originally planned to last 15 years. However, rapid technological advances and worldwide participation have accelerated the expected completion date to 2003. In June 2000, scientists announced the generation of a working draft sequence of the entire human genome. The draft provides a road map to an estimated 90% of genes on every human chromosome. Already it has enabled gene hunters to pinpoint genes associated with more than 30 disorders.

Biotechnological engineering or biological engineering is a branch of engineering that focuses on biotechnologies and biological science. It includes different disciplines such as biochemical engineering, biomedical engineering, bio-process engineering, biosystem engineering and so on. Because of the novelty field, the definition of a bioengineer is still ill defined. However, in general it is an integrated approach of fundamental biological sciences and traditional engineering principles. Bioengineers are often employed to scale up a bio processes from the laboratory scale to the manufacturing scale. Moreover, as with most engineers, they often deal with management, economic and legal issues. Since patents and regulation (e.g. FDA regulation in the U.S.) are very important issues for biotech enterprises, bioengineers are often required to have knowledge related to these issues. The increasing number of biotech enterprises is likely to create a need for bioengineers in the years to come. Many universities throughout the world are now providing programs in bioengineering and biotechnology.

Human cloning is one of the techniques of modern biotechnology. It involves the removal of the nucleus from one cell and its placement in an unfertilized egg cell whose nucleus has either been deactivated or removed.

There are three types of cloning:

- Reproductive cloning. After a few divisions, the egg cell is placed into a uterus where it is allowed to develop into a fetus that is genetically identical to the donor of the original nucleus.
- Therapeutic cloning. The egg is placed into a Petri dish where it develops into embryonic stem cells, which have shown potentials for treating several ailments.

In February 1997, cloning became the focus of media attention when Ian Wilmut and his colleagues at the Roslin Institute announced the successful cloning of a sheep, named Dolly, from the mammary glands of an adult female. The cloning of Dolly made it apparent to many that the techniques used to produce her could someday be used to clone human beings. This stirred a lot of controversy because of its ethical implications.

Several issues have been raised regarding the use of modern biotechnology in the medical sector. Many of these issues are similar to those facing any new technology that is viewed as powerful and far-reaching. Some of these issues are:

1. Absence of cure. There is still a lack of effective treatment or preventive measures for many diseases and conditions now being diagnosed or predicted using gene tests. Thus, revealing information about risk of a future disease that has no existing cure presents an ethical dilemma for medical practitioners.
2. Ownership and control of genetic information. Who will own and control genetic information, or information about genes, gene products, or inherited characteristics derived from an individual or a group of people like indigenous communities? At the macro level, there is a possibility of a genetic divide, with developing countries that do not have access to medical applications of biotechnology being deprived of benefits accruing from products derived from genes obtained from their own people. Moreover, genetic information can pose a risk for minority population groups as it can lead to group stigmatisation. At the individual level, the absence of privacy and anti-discrimination legal protections in most countries can lead to discrimination in employment or insurance or other misuse of personal genetic information. This raises questions like, is genetic privacy different from medical privacy?
3. Reproductive issues. These include the use of genetic information in reproductive decision-making and the possibility of genetically altering reproductive cells that may be passed on to future generations. For example, germline therapy forever changes the genetic make-up of an individual's descendants. Thus, any error in technology or judgment may have far-reaching consequences. Ethical issues like designer babies and human cloning have also given rise to controversies between and among scientists and bioethicists, especially in the light of past abuses with eugenics.
4. Clinical issues. These center on the capabilities and limitations of doctors and other health-service providers, people identified with genetic conditions, and the general public in dealing with genetic information. For instance, how should the public be prepared to make informed choices based on the results of genetic tests? How will genetic tests be evaluated and regulated for accuracy, reliability, and usefulness?

5. Effects on social institutions. Genetic tests reveal information about individuals and their families. Thus, test results can affect the dynamics within social institutions, particularly the family.
6. Conceptual and philosophical implications regarding human responsibility, free will vis-à-vis genetic determinism, and the concepts of health and disease. Do genes influence human behavior? If so, does genetic testing mean controlling human behavior? What is considered acceptable diversity? What is normal and what is a disability or disorder, and who decides these matters? Are disabilities diseases that need to be cured or prevented? Where should the line between medical treatment and enhancement be drawn? Who will have access to gene therapy?

In its broadest sense, "biotechnology" refers to "any technique that uses living organisms, or parts of such organisms, to make or modify products, to improve plants or animals, or to develop microorganisms for specific use." Biotechnology has evolved through the years.

On one end of the development pole are techniques of traditional biotechnology like microbial fermentation, used as early as 10,000 years ago in fermenting beer, wine and dairy products. At the other end of the development pole are the continuously evolving techniques of modern biotechnology, such as genetic engineering. Using genetic engineering techniques, the genetic makeup of an organism may be modified by inactivating or altering some of its genes and introducing other natural or artificial genes, usually from another organism.

The Cartagena Protocol on Biosafety defines modern biotechnology as referring to any process that involves the application of (i) *in vitro* nucleic acid techniques, including recombinant deoxyribonucleic acid and direct injection of nucleic acid into cells or organelles, or (ii) fusion of cells beyond the taxonomic family, that overcome natural physiological reproductive or recombination of barriers and that are not techniques used in traditional breeding and selection.

Although the Protocol is not yet in force (because less than the required 50 States have either ratified or acceded to it), the Protocol's definition of modern biotechnology has gained currency in international circles. However, while there may be an emerging international consensus on the above definition, strictly speaking it is a definition that is applicable only when one uses the term "modern biotechnology" for purposes of interpreting or implementing the Protocol.

Green Revolution Through Biotechnology

The world's worst recorded food disaster occurred in 1943 in British-ruled India. Known as the Bengal Famine, an estimated 4 million people died of hunger that year in eastern India (which included today's Bangladesh). Initially, this catastrophe was attributed to an acute shortfall in food production in the area. However, Indian economist Amartya Sen (recipient of the Nobel Prize for Economics, 1998) has established that while food shortage was a contributor to the problem, a more potent factor was the result of hysteria related to World War II, which made food supply a low priority for the British rulers.

When the British left India in 1947, India continued to be haunted by memories of the Bengal Famine. It was therefore natural that food security was one of the main items on free India's agenda. This awareness led, on one hand, to the Green Revolution in India and, on the other, legislative measures to ensure that businessmen would never again be able to hoard food for reasons of profit.

The Green Revolution, spreading over the period from1967/68 to 1977/78, changed India's status from a food-deficient country to one of the world's leading agricultural nations. Until 1967 the government largely concentrated on expanding the farming areas. But the population was growing at a much faster rate than food production. This called for an immediate and drastic action to increase yield. The action came in the form of the Green Revolution. The term 'Green Revolution' is a general one that is applied to successful agricultural experiments in many developing countries. India is one of the countries where it was most successful.

There were three basic elements in the method of the Green Revolution

- Continuing expansion of farming areas
- Double-cropping in the existing farmland
- Using seeds with improved genetics.

The area of land under cultivation was being increased from 1947 itself. But this was not enough to meet the rising demand. Though other methods were required, the expansion of cultivable land also had to continue. So, the Green Revolution continued with this quantitative expansion of farmlands.

Double cropping was a primary feature of the Green Revolution. Instead of one crop season per year, the decision was made to have two crop seasons per year. The one-season-per-year practice was based on the fact that there is only one rainy season annually. Water for the second phase now came from huge irrigation projects. Dams were built and other simple irrigation techniques were also adopted.

Using seeds with superior genetics was the scientific aspect of the Green Revolution. The Indian Council for Agricultural Research (which was established by the British in 1929) was reorganised in 1965 and then again in 1973. It developed new strains of high yield variety seeds, mainly wheat and rice and also millet and corn.

The Green Revolution was a technology package comprising material components of improved high yielding varieties of two staple cereals (rice and wheat), irrigation or controlled water supply and improved moisture utilisation, fertilizers, and pesticides, and associated management skills.

Shortcomings

In spite of this, India's agricultural output sometimes falls short of demand even today. India has failed to extend the concept of high yield value seeds to all crops or all regions. In terms of crops, it remains largely confined to foodgrains only, not to all kinds of agricultural produce.

In regional terms, only the states of Punjab and Haryana showed the best results of the Green Revolution. The eastern plains of the River Ganges in West Bengal also showed reasonably good results. But results were less impressive in other parts of India.

The Green Revolution has created some problems mainly to adverse impacts on the environment. The increasing use of agrochemical-based pest and weed control in some crops has affected the surrounding environment as well as human health. Increase in the area under irrigation has led to rise in the salinity of the land. Although high yielding varieties had their plus points, it has led to significant genetic erosion.

Merits

Thanks to the new seeds, tens of millions of extra tonnes of grain a year are being harvested. The Green Revolution resulted in a record grain output of 131 million tonnes in 1978/79. This established India as one of the world's biggest agricultural producers. Yield per unit of farmland improved by more than 30% between1947 (when India gained political independence) and 1979. The crop area under high yielding varieties of wheat and rice grew considerably during the Green Revolution. The Green Revolution also created plenty of jobs not only for agricultural workers but also industrial workers by the creation of related facilities such as factories and hydroelectric power stations.

APPENDICES

Appendix - I

TIMELINE OF BIOLOGY AND ORGANIC CHEMISTRY

Before 1600

- c. 500 B.C. - Alcmaeon of Croton distinguished veins from arteries and discovered the optic nerve.
- c. 500 B.C. - Sushruta - wrote Sushruta Samhita describing over 120 surgical instruments, 300 surgical procedures and classified human surgery in 8 categories. Performed cosmetic surgery.
- c. 500 B.C. - Xenophanes examined fossils and speculated on the evolution of life.
- c. 350 B.C. - Aristotle attempted a comprehensive classification of animals. His written works include *Historia Animalium*, a general biology of animals, *De Partibus Animalium*, a comparative anatomy and physiology of animals, and *De Generatione Animalium*, on developmental biology.
- c. 320 BC - Theophrastos (or Theophrastus) begins the systematic study of botany.
- c. 300 B.C. - Herophilos dissected the human body.
- c. 300 B.C. - Diocles wrote the first known anatomy book and was the first to use the term *anatomy*.
- c. 50-70 - *Historia Naturalis* by Pliny the Elder (Gaius Plinius Secundus) was published in 37 volumes.
- 130-200 - Claudius Galen wrote numerous treatises on human anatomy.
- c. 1010 - Avicenna (Ibn Sina or Abu Ali al Hussein ibn Abdallah) published his *Canon of Medicine* (Kitab al-Qanun fi al-tibb).
- 1543 - Andreas Vesalius publishes the anatomy treatise *De humani corporis fabrica*.

1600-1699

- 1628 - William Harvey publishes *An Anatomical Exercise on the Motion of the Heart and Blood in Animals*
- 1651 - William Harvey concludes that all animals, including mammals, develop from eggs, and spontaneous generation of any animal from mud or excrement was an impossibility.
- 1658 - Jan Swammerdam observes red blood cells under a microscope.
- 1663 - Robert Hooke sees cells in cork using a microscope.
- 1668 - Francesco Redi disproves spontaneous generation by showing that fly maggots only appear on pieces of meat in jars if the jars are open to the air. Jars covered with cheesecloth contained no flies.
- 1672 - Marcello Malpighi publishes the first description of chick development, including the formation of muscle somites, circulation, and nervous system.
- 1676 - Anton van Leeuwenhoek observes protozoa and calls them *animalcules*.
- 1677 - Anton van Leeuwenhoek observes spermatozoa.
- 1683 - Anton van Leeuwenhoek observes bacteria. Leeuwenhoek's discoveries renew the question of spontaneous generation in microorganisms.

1700-1799

- 1767 - Kaspar Friedrich Wolff argues that the tissues of a developing chick form from nothing and are not simply elaborations of already-present structures in the egg.
- 1768 - Lazzaro Spallanzani again disproves spontaneous generation by showing that no organisms grow in a rich broth if it is first heated (to kill any organisms) and allowed to cool in a stoppered flask. He also shows that fertilization in mammals requires an egg and semen.
- 1771 - Joseph Priestley demonstrates that plants produce a gas that animals and flames consume. Those two gases are carbon dioxide and oxygen.
- 1798 - Thomas Malthus discusses human population growth and food production in *An Essay on the Principle of Population*.

1800-1899

- 1801 - Jean-Baptiste Lamarck begins the detailed study of invertebrate taxonomy.
- 1802 - The term *biology* in its modern sense is propounded independently by Gottfried Reinhold Treviranus (*Biologie oder Philosophie der lebenden Natur*)

and Lamarck (*Hydrogéologie*). The word had been coined in 1800 by Karl Friedrich Burdach.

- 1809 - Lamarck proposes a modern theory of evolution based on the inheritance of acquired characteristics.
- 1817 - Pierre-Joseph Pelletier and Joseph-Bienaime Caventou isolate chlorophyll.
- 1820 - Christian Friedrich Nasse formulates Nasse's law: hemophilia occurs only in males and is passed on by unaffected females.
- 1824 - J. L Prevost and J. B. Dumas showed that the sperm in semen were not parasites, as previously thought, but, instead, the agents of fertilization.
- 1826 - Karl von Baer shows that the eggs of mammals are in the ovaries, ending a 200-year search for the mammalian egg.
- 1828 - Friedrich Woehler synthesizes urea; first synthesis of an organic compound from inorganic starting materials.
- 1836 - Theodor Schwann discovers pepsin in extracts from the stomach lining; first isolation of an animal enzyme.
- 1837 - Theodor Schwann shows that heating air will prevent it from causing putrefaction.
- 1838 - Matthias Schleiden proposes that all plants are composed of cells.
- 1839 - Theodor Schwann proposes that all animal tissues are composed of cells. Schwann and Schleinden argued that cells are the elementary particles of life.
- 1856 - Louis Pasteur states that microorganisms produce fermentation.
- 1858 - Charles R. Darwin and Alfred Wallace independently propose a theory of biological evolution ("descent through modification") by means of natural selection. Only in later editions of his works did Darwin used the term "evolution."
- 1858 - Rudolf Virchow proposes that cells can only arise from pre-existing cells; "Omnis cellula e celulla," all cell from cells. The Cell Theory states that all organisms are composed of cells (Schleiden and Schwann), and cells can only come from other cells (Virchow).
- 1864 - Louis Pasteur disproves the spontaneous generation of cellular life.
- 1865 - Gregor Mendel demonstrates in pea plants that inheritance follows definite rules. The Principle of Segregation states that each organism has two genes per trait, which segregate when the organism makes eggs or sperm. The Principle of Independent Assortment states that each gene in a

pair is distributed independently during the formation of eggs or sperm. Mendel's trailblazing foundation for the science of genetics went unnoticed, to his lasting disappointment.

- 1865 - Friedrich August Kekulé von Stradonitz realizes that benzene is composed of carbon and hydrogen atoms in a hexagonal ring.
- 1869 - Friedrich Miescher discovers nucleic acids in the nuclei of cells.
- 1874 - Jacobus van 't Hoff and Joseph-Achille Le Bel advance a three-dimensional stereochemical representation of organic molecules and propose a tetrahedral carbon atom.
- 1876 - Oskar Hertwig and Hermann Fol independently describe (in sea urchin eggs) the entry of sperm into the egg and the subsequent fusion of the egg and sperm nuclei to form a single new nucleus.
- 1884 - Emil Fischer begins his detailed analysis of the compositions and structures of sugars.
- 1892 - Hans Driesch separates the individual cells of a 2-cell sea urchin embryo and shows that each cell develops into a complete individual, thus disproving the theory of preformation and showing that each cell is "totipotent," containing all the hereditary information necessary to form an individual.
- 1898 - Martinus Beijerinck uses filtering experiments to show that tobacco mosaic disease is caused by something smaller than a bacterium, which he names a virus.

1900-1949

- 1900 - Two biologists independently rediscover Mendel's paper on heredity.
- 1902 - Walter Sutton and Theodor Boveri, independently propose that the chromosomes carry the hereditary information.
- 1905 - William Bateson coins the term "genetics" to describe the study of biological inheritance.
- 1906 - Mikhail Tsvet discovers the chromatography technique for organic compound separation.
- 1907 - Ivan Pavlov demonstrates conditioned responses with salivating dogs.
- 1907 - Emil Fischer artificially synthesizes peptide amino acid chains and thereby shows that amino acids in proteins are connected by amino group-acid group bonds.
- 1909 - Wilhelm Johannsen coined the word "gene."

- 1911 - Thomas Hunt Morgan proposes that genes are arranged in a line on the chromosomes.
- 1926 - James Sumner shows that the urease enzyme is a protein.
- 1928 - Otto Diels and Kurt Alder discover the Diels-Alder cycloaddition reaction for forming ring molecules.
- 1928 - First antibiotic, penicillin, discovered by Alexander Fleming
- 1929 - Phoebus Levene discovers the sugar deoxyribose in nucleic acids.
- 1929 - Edward Doisy and Adolf Butenandt independently discover estrone.
- 1930 - John Howard Northrop shows that the pepsin enzyme is a protein.
- 1931 - Adolf Butenandt discovers androsterone.
- 1932 - Hans Adolf Krebs discovers the urea cycle.
- 1933 - Tadeus Reichstein artificially synthesizes vitamin C; first vitamin synthesis.
- 1935 - Rudolf Schoenheimer uses deuterium as a tracer to examine the fat storage system of rats.
- 1935 - Wendell Stanley crystallizes the tobacco mosaic virus.
- 1935 - Konrad Lorenz describes the imprinting behavior of young birds.
- 1937 - Dorothy Crowfoot Hodgkin discovers the three-dimensional structure of cholesterol.
- 1937 - Hans Adolf Krebs discovers the tricarboxylic acid cycle.
- 1937 - In Genetics and the Origin of Species, Theodosius Dobzhansky applies the chromosome theory and population genetics to natural populations in the first mature work of neo-Darwinism, also called the modern synthesis, a term coined by Julian Huxley.
- 1938 - A living coelacanth is found off the coast of southern Africa.
- 1940 - Donald Griffin and Robert Galambos announce their discovery of sonar echolocation by bats.
- 1942 - Max Delbruck and Salvador Luria demonstrate that bacterial resistance to virus infection is caused by random mutation and not adaptive change.
- 1944 - Oswald Avery shows that DNA carries the genetic code in pneumococcus bacteria.
- 1944 - Robert Burns Woodward and William von Eggers Doering synthesize quinine.

- 1945 - Dorothy Crowfoot Hodgkin discovers the three-dimensional structure of penicillin.
- 1948 - Erwin Chargaff shows that in DNA the number of guanine units equals the number of cytosine units and the number of adenine units equals the number of thymine units.

1950-1989

- 1951 - Robert Woodward synthesizes cholesterol and cortisone.
- 1952 - American developmental biologists Robert Briggs and Thomas King clone the first vertebrate by transplanting nuclei from leopard frogs embryos into enucleated eggs. More differentiated cells were the less able they are to direct development in the enucleated egg.
- 1952 - Alfred Hershey and Martha Chase show that DNA is the genetic material in bacteriophage viruses.
- 1952 - Fred Sanger, Hans Tuppy, and Ted Thompson complete their chromatographic analysis of the insulin amino acid sequence.
- 1952 - Rosalind Franklin concludes that DNA is a double helix with a diameter of 2 nm and the sugar-phosphate backbones on the outside of the helix, based on x ray diffraction studies. She suspects the two sugar-phosphate backbones have a peculiar relationship to each other.
- 1953 - After examining Franklin's unpublished data, James D. Watson and Francis Crick publish a double-helix structure for DNA, with one sugar-phosphate backbone running in the opposite direction to the other. They further suggest a mechanism by which the molecule can replicate itself and serve to transmit genetic information. Their paper, combined with the Hershey-Chase experiment and Chargaff's data on nucleotides, finally persuades biologists that DNA is the genetic material, not protein.
- 1953 - Max Perutz and John Kendrew determine the structure of hemoglobin using X-ray diffraction studies.
- 1953 - Stanley Miller shows that amino acids can be formed when simulated lightning is passed through vessels containing water, methane, ammonia, and hydrogen
- 1954 - Dorothy Crowfoot Hodgkin discovers the three-dimensional structure of vitamin B-12.
- 1955 - Marianne Grunberg-Manago and Severo Ochoa discover the first nucleic-acid-synthesizing enzyme (polynucleotide phosphorylase), which links nucleotides together into polynucleotides.

- 1955 - Arthur Kornberg discovers DNA polymerase enzymes.
- 1958 - Matthew Stanley Meselson and Franklin W. Stahl prove that DNA replication is semiconservative in the Meselson-Stahl experiment
- 1959 - Severo Ochoa and Arthur Kornberg receive a Nobel Prize for their work.
- 1959 - Max Perutz describes the structure of hemoglobin, the oxygen-carrying protein in blood.
- 1960 - John Kendrew describes the structure of myoglobin, the oxygen-carrying protein in muscle.
- 1960 - Four separate researchers (S. Weiss, J. Hurwitz, Audrey Stevens and J. Bonner) discover bacterial RNA polymerase, which polymerizes nucleotides under the direction of DNA.
- 1960 - Juan Oro finds that concentrated solutions of ammonium cyanide in water can produce the nucleotide organic base adenine.
- 1960 - Robert Woodward synthesizes chlorophyll.
- 1961 - German plant physiologist H. J. Matthaei cracks the first codon of the genetic code (the codon for the amino acid phenylalanine) using Grunberg-Manago's enzyme system for making polynucleotides.
- 1962 - Max Perutz and John Kendrew share a Nobel prize for their work on the structure of hemoglobin and myoglobin.
- 1965 - Genetic code fully cracked through trial-and-error experimental work.
- 1966 - Kimishige Ishizaka discovers a new type of immunoglobulin, IgE, that develops allergy and explains the mechanisms of allergy at molecular and cellular levels.
- 1967 - John Gurden uses nuclear transplantation to clone an African clawed frog; first cloning of a vertebrate using a nucleus from a fully differentiated adult cell.
- 1968 - Fred Sanger uses radioactive phosphorus as a tracer to chromatographically decipher a 120 base long RNA sequence.
- 1969 - Dorothy Crowfoot Hodgkin discovers the three-dimensional structure of insulin.
- 1970 - Hamilton Smith and Daniel Nathans discover DNA restriction enzymes.
- 1970 - Howard Temin and David Baltimore independently discover reverse transcriptase enzymes.

- 1972 - Robert Woodward synthesizes vitamin B-12.
- 1972 - Stephen Jay Gould and Niles Eldredge propose an idea they call "punctuated equilibrium," which states that the fossil record is an accurate depiction of the pace of evolution, with long periods of "stasis" (little change) punctuated by brief periods of rapid change and species formation (within a lineage).
- 1972 - SJ Singer and GL Nicholson develop the fluid mosaic model, which deals with the make-up of the membrane of all cells.
- 1974 - Manfred Eigen and Manfred Sumper show that mixtures of nucleotide monomers and RNA replicase will give rise to RNA molecules which replicate, mutate, and evolve.
- 1974 - Leslie Orgel shows that RNA can replicate without RNA-replicase and that zinc aids this replication.
- 1977 - John Corliss, Jack Dymond, Louis Gordon, John Edmond, Richard von Herzen, Robert Ballard, Kenneth Green, David Williams, Arnold Bainbridge, Kathy Crane, and Tjeerd van Andel discover chemosynthetically based animal communities located around submarine hydrothermal vents on the Galapagos Rift.
- 1977 - Walter Gilbert and Allan Maxam present a rapid DNA sequencing technique which uses cloning, base destroying chemicals, and gel electrophoresis.
- 1977 - Frederick Sanger and Alan Coulson present a rapid gene sequencing technique which uses dideoxynucleotides and gel electrophoresis.
- 1978 - Frederick Sanger presents the 5,386 base sequence for the virus PhiX174; first sequencing of an entire genome.
- 1982 - Stanley B. Prusiner proposes the existence of infectious proteins, or prions. His idea is widely derided in the scientific community, but he wins a Nobel Prize in 1997.
- 1983 - Kary Mullis invents "PCR" (polymerase chain reaction), an automated method for rapidly copying sequences of DNA.
- 1984 - Alec Jeffreys devises a genetic fingerprinting method.
- 1985 - Harry Kroto, J.R. Heath, S.C. O'Brien, R.F. Curl, and Richard Smalley discover the unusual stability of the buckminsterfullerene molecule and deduce its structure.
- 1986 - Alexander Klibanov demonstrates that enzymes can function in non-aqueous environments.

1990-Present

- 1990 - Napoli, Lemieux and Jorgensen discover RNA interference (1990) during experiments aimed at the color of petunias.
- 1990 - Wolfgang Krätschmer, Lowell Lamb, Konstantinos Fostiropoulos, and Donald Huffman discover that Buckminsterfullerene can be separated from soot because it is soluble in benzene.
- 1995 - Publication of the first complete genome of a free-living organism.
- 1996 - Dolly the sheep is first clone of an adult mammal.
- 2001 - Publication of the first drafts of the complete human genome.
- 2002 - First virus produced 'from scratch,' an artificial polio virus that paralyzes and kills mice.

BIOGAS

Biogas typically refers to a (biofuel) gas produced by the anaerobic digestion or fermentation of organic matter including manure, sewage sludge, municipal solid waste, biodegradable waste or any other biodegradable feedstock, under anaerobic conditions. Biogas is comprised primarily of methane and carbon dioxide.

Depending on where it is produced, biogas is also called:

- swamp gas
- marsh gas
- landfill gas
- digester gas

Biogas containing methane is a valuable by-product of anaerobic digestion which can be utilised in the production of renewable energy.Biogas can be used as a vehicle fuel or for generating electricity. It can also be burned directly for cooking, heating, lighting, process heat and absorption refrigeration.

Biogas and Anaerobic Digestion

Biogas production by anaerobic digestion is popular for treating biodegradable waste because valuable fuel can be produced, while destroying disease-causing pathogens and reducing the volume of disposed waste products. It burns more cleanly than coal, and emits less carbon dioxide per unit of energy. The harvesting of biogas is an important part of waste management because methane is a greenhouse gas with a greater global warming potential than carbon dioxide. The carbon in biogas was generally recently extracted from the atmosphere by photosynthetic plants, so releasing it back into the atmosphere adds less total atmospheric carbon than burning fossil fuels.

Recently, developed countries have been making increasing use of biogas generated from both wastewater and landfill sites or produced by mechanical biological treatment systems for municipal waste. High energy prices and increases in subsidies for electricity from renewable sources (such as renewables obligation certificates) and drivers such as the EU Landfill Directive have led to much greater use of biogas sources.

Landfill Gas

Landfill gas is produced from organic waste disposed of in landfill. The waste is covered and compressed mechanically and by the pressure of higher levels. As conditions become anaerobic the organic waste is broken down and landfill gas is produced. This gas builds up and is slowly released into the atmosphere. This is hazardous for three key reasons:

- Risk of explosion
- Global warming through methane as a greenhouse gas
- Volatile organic compounds (VOCs) as precursor to photochemical smog

Biogas Composition

The composition of biogas varies depending upon the origin of the anaerobic digestion process. Landfill gas typically has methane concentrations around 50%. Advanced waste treatment technologies can produce biogas with 55-75%CH4.

*often 5 % of air is introduced for microbiological desulphurisation

(GWh) 17 272 13 397

Siloxanes and Gas Engines *EU*

Typical Composition of Biogas

Matter	%
Methane, CH_4	50-75
Carbon dioxide, CO_2	25-50
Nitrogen, N_2	0-10
Hydrogen, H_2	0-1
Hydrogen sulphide, H_2S	0-3
Oxygen, O_2	0-2

Biogas in EU 2006 (GWh)

Country	*Total*	*Landfill*	*Sludge*	*Other*
Germany	22370	66704	300	11,400
UK	19720	17,620	2100	0
Italy	4110	3610	10	490
Spain	3890	2930	660	300
France	2640	1720	870	50
Netherlands	1380	450	590	340
Austria	1370	130	40	1 200
Denmark	1100	170	270	660
Poland	1090	320	770	10
Belgium	970	590	290	90
Greece	810	630	180	0
Finland	740	590	150	0
Czech Republic	700	300	360	40
Ireland	400	290	60	50
Sweden	390	130	250	10
Hungary	120	0	90	40
Portugal	110	0	0	110
Luxembourg	100	0	0	100
Slovenia	100	80	10	10
Slovakia	60	0	50	10
Estonia	10	10	0	0
Malta	0	0	0	0
EU (GWh)	62 200	36 250	11 050	14 900

Electricity from Biogas (GWh)

Country	*2006*	*2005*
Germany	7338	4708
UK	4997	4690
Italy	1234	1198
Spain	675	620
Greece	579	179
France	501	483
Austria	410	70
Netherlands	286	286
Denmark	285	275
Poland	241	175
Belgium	237	240
Czech Republic	175	161
Ireland	108	106
Sweden	54	54
Portugal	33	35
Luxembourg	33	27
Slovenia	32	32
Hungary	22	25

Finland	22	22
Estonia	7	7
Slovakia	4	4
Malta	0	0
EU (GWh)	17 272	13 39

In some cases, biogas from landfills and sewage treatment contains siloxanes. During combustion of biogas containing siloxanes, silicon is released and can combine with free oxygen or various other elements in the combustion gas. Deposits are formed containing mostly silica (SiO_2) or silicates (Si_xO_y) in general, but can also contain calcium, sulphur, zinc, phosphor... as indicated by the analysis piston scrapings from biogas-fired engines. These (mostly white) deposits can ultimately build to a surface thickness of several millimetres and are difficult to remove by chemical or mechanical means. In internal combustion engines deposits on pistons and cylinder heads are extremely abrasive and even a small amount is sufficient to cause enough damage to the engine to require a complete overhaul at 5,000 h or less of operation. The damage is similar to that caused by carbon build up during light load running of diesel engines. Deposits on the turbine of the turbocharger will eventually reduce the charger's efficiency. Luckily, simply cooling the gas to roughly -4 C is sufficient to remove siloxanes due to condensantion. Stirling engines are more resistant against siloxanes, though deposits on the tubes of the heat exchanger will reduce the efficiency.

Safety

The methane in biogas forms explosive mixtures in air. The lower explosive limit is 5% methane and the upper explosive limit is 15% methane.

Biogas to Natural Gas

If biogas is cleaned up sufficiently, biogas has the same characteristics as natural gas. In this instance the producer of the biogas can utilize the local gas distribution networks. The gas must be very clean to reach pipeline quality. Water (H_2O), hydrogen sulfide (H_2S) and particulates are removed if present at high levels or if the gas is to be completely cleaned. Carbon dioxide is less frequently removed, but it must also be separated to achieve pipeline quality gas. If the gas is to be used without extensively cleaning, it is sometimes cofired with natural gas to improve combustion. Biogas cleaned up to pipeline quality is called renewable natural gas or biomethane.

Applications of Renewable Natural Gas

In this form the gas can be now used in any application that natural gas is used for. Such applications include distribution via the natural gas grid, electricity production, space heating, water heating and process heating. If compressed, it can replace compressed natural gas for use in vehicles, where it can fuel an internal combustion engine or fuel cells.

Cooking

Gober gas is a biogas generated out of cow dung. In India, gober gas is generated at the countless number of micro plants (an estimated more than 2 million) attached to households. The gober gas plant is basically an airtight circular pit made of concrete with a pipe connection. The manure is directed to the pit (usually directed from the cattle shed). The pit is then filled with a required quantity of water (usually waste water). The gas pipe is connected to the kitchen fire place through control valves. The flammable methane gas generated out of this is practically odorless and smokeless. The residue left after the extraction of the gas is used as biofertiliser. Owing to its simplicity in implementation and use of cheap raw materials in the villages, it is often quoted as one of the most environmentally sound energy source for the rural needs.

Railway Transport

A biogas-powered train has been in service in Sweden since 2005.

Landfill Gas Legislation

United States

In the United States, because landfill gas contains these VOCs, the United States Clean Air Act and Title 40 of the Code of Federal Regulations (CFR) requires landfill owners to estimate the quantity of non-methane organic compounds (NMOCs) emitted. If the estimated NMOC emissions exceeds 50 tonnes per year the landfill owner is required to collect the landfill gas and treat it to remove the entrained NMOCs. Treatment of the landfill gas is usually by combustion. Because of the remoteness of landfill sites it is sometimes not economically feasible to produce electricity from the gas.

BIOFUEL

Biofuel (also called agrofuel) can be broadly defined as solid, liquid, or gas fuel consisting of, or derived from biomass. This article, however, is principally about biofuel in the form of liquid or gas transportation fuel derived from biomass. Biomass can also be used directly for heating or power. One type of biomass is wood, which is frequently used in industry, either by itself to create energy or with other combustible matter (such as coal) to burn and create heat. (Wood has been burned for millennia - as solids.)

Biofuel is considered a means of reducing greenhouse gas emissions and increasing energy security by providing an alternative to fossil fuels. However, In October 2007, Nobel Laureate Paul Crutzen published findings that the release of Nitrous Oxide (N2O) among the commonly used biofuels, such as biodiesel from rapeseed and bioethanol from corn (maize), can contribute as much or more to global warming than fossil fuel savings due to global cooling. Crops with less N demand, such as grasses and woody coppice species have more favourable climate impacts. Biofuels are used globally: biofuel industries are expanding in Europe, Asia and the Americas. The most common use for biofuels is in automotive transport (for example E10 fuel). Biofuel can be produced from any carbon source that can be replenished rapidly e.g. plants. Many different plants and plant-derived materials are used for biofuel manufacture.

Biomass

Biomass is material derived from recently living organisms. This includes plants, animals and their by-products. For example, manure, garden waste and crop residues are all sources of biomass. It is a renewable energy source based on the carbon cycle, unlike other natural resources such as petroleum, coal, and nuclear fuels. Agricultural products specifically grown for biofuel production include corn, switchgrass, and soybeans, primarily in the United States; rapeseed, wheat and sugar beet primarily in Europe; sugar cane in Brazil; palm oil and miscanthus in South-East Asia; sorghum and cassava in China; and jatropha in India. Hemp has also been proven to work as a biofuel. Biodegradable outputs from industry, agriculture, forestry and households can be used for biofuel production, either using anaerobic digestion to produce biogas, or using second generation biofuel processes; examples include straw, timber, manure, rice husks, sewage, and food waste. The use of biomass fuels can therefore contribute to waste management as well as fuel security and help to prevent climate change, though alone they are not a comprehensive solution to these problems.

History

Humans have used biomass fuels - that is, solid biofuels - for heating and cooking since the discovery of fire. Following the discovery of electricity, it became possible to use biofuels to generate electrical power as well. However, the discovery and use of fossil fuels: coal, gas and oil, have dramatically reduced the amount of biomass fuel used in the developed world for transport, heat and power. Liquid biofuels have been used since the early days of the car industry. Nikolaus August Otto, the German inventor of the internal combustion engine, conceived his invention to run on ethanol. Rudolf Diesel, the German inventor of the Diesel engine, designed it to run on peanut oil. Henry Ford originally designed the Ford Model T, a car produced from 1903 to 1926, to run completely on ethanol. However, when crude oil became cheaply available (thanks to oil reserves discovered in Pennsylvania and Texas), cars began using fuels derived from mineral oil: petroleum or diesel. Nevertheless, before World War II, biofuels were seen as providing an alternative to imported oil. Germany powered its vehicles using a blend of gasoline with alcohol fermented from potatoes, called Reichskraftsprit. In Britain, grain alcohol was blended with petrol by the Distillers Company Limited under the name Discol and marketed through Esso's affiliate Cleveland. After the war, cheap Middle Eastern oil lessened interest in biofuels. But the oil shocks of 1973 and 1979 increased interest from governments and academics. The counter-shock of 1996 again reduced oil prices and interest. In the United States, all cars manufactured since 1988 are required to be compatible with fuels containing at least 20% ethanol E20 fuel, and with minor modifications these cars can use 85% ethanol blended with petroleum E85 fuel. Since around 2000 renewed interest in biofuels has been seen. The drivers for biofuel use and development include rising oil prices, concerns over the potential oil peak, greenhouse gas emissions (global warming), rural development interests, and instability in the Middle East. The US president George W. Bush said in his 2006 State of the Union speech that the US should replace 75% of imported oil with biofuel by 2025. The U.S. Dept. of Energy has earmarked $375 million to fund bioenergy research centers. Second generation biofuel production processes are in development. These allow biofuel to be derived from any source of biomass, not just from food crops such as corn and soy beans.

Carbon Emissions

Biofuels and other forms of renewable energy aim to be carbon neutral. This means that the carbon released during the use of the fuel, e.g. through burning to power transport or generate electricity, is reabsorbed and balanced by the carbon absorbed by new plant growth. These plants are then harvested to make the next batch of fuel. Carbon neutral fuels lead to no net increases in atmospheric carbon dioxide levels, which means that global warming need not get any worse. In practice, biofuels are not carbon neutral. This is because energy is required to grow crops and process them into fuel. Examples of energy use during the production of biofuels include: fertilizer manufacture, fuel used to power machinery, and fuel used to transport crops and fuels to and from biofuel processing plants. The amount of fuel used during biofuel production has a large impact on the overall greenhouse gas emissions savings achieved by biofuels. In October 2007, further doubt has been thrown on the advantages of biofuel by Nobel Laureate Paul Crutzen, who says that the advantages of reduced carbon dioxide emissions are more than offset by increased nitrous oxide emissions. Nitrous oxide is both a potent greenhouse gas and a destroyer of atmospheric ozone.

The carbon emissions produced by biofuels are calculated using a technique called Life Cycle Analysis (LCA). This uses a "cradle to grave" or "well to wheels" approach to calculate the total amount of carbon dioxide and other greenhouse gases emitted during biofuel production, from putting seed in the ground to using the fuel in cars and trucks. Many different LCAs have been done for different biofuels, with widely differing results. The majority of LCA studies show that biofuels provide significant greenhouse gas emissions savings when compared to fossil fuels such as petroleum and diesel. Therefore, using biofuels to replace a proportion of the fossil fuels that are burned for transportation can reduce overall greenhouse gas emissions. This does assume however that the land used for growing the crops would alternatively be desert or paved area. If the land was previously a (tropical rain-) forest, the carbon absorption of this forest should be deducted from the greenhouse gas savings. This implies that the net effect of burning bio-fuels is an increase in greenhouse gasses. This effect should be incorporated in the LCA, to get a proper overview of the total net effect. Using waste material from plantation forests on previous agricultural land could be carbon positive, due to the carbon stored below ground in the root systems. The well-to-wheel analysis for biofuels has shown that first generation biofuels can save up to 60% carbon emission and second generation biofuels can save up to 80% as opposed to using fossil fuels. However, a 2007 study by scientists from Britain, U.S., Germany, Switzerland and including Professor Paul Crutzen, who won a Nobel Prize for his work on ozone, have reported that measurements of emissions from the burning of biofuels derived from rapeseed and corn have been found to produce more greenhouse gas emissions than they save. The claim that biofuels result in emissions savings has also been critiqued on the grounds that it overlooks the 'displacement' effects of large-scale biofuel production, in terms of its direct and indirect role in promoting land use changes and soil carbon losses. In 2006, a UK Government study showed that carbon emissions were reduced between 50% and 60%. This was when biofuels were used in conjunction with other fuels such as petrol and diesel.

Bioenergy from Waste

Using waste biomass to produce energy can reduce the use of fossil fuels, reduce greenhouse gas emissions and reduce pollution and waste management problems. A recent publication by the European Union highlighted the potential for waste-derived bioenergy to contribute to the reduction of global warming. The report concluded that 19 million tons of oil equivalent is available from biomass by 2020, 46% from bio-wastes: municipal solid waste (MSW), agricultural residues, farm waste and other biodegradable waste streams. Landfill sites generate gases as the waste buried in them undergoes anaerobic digestion. These gases are known collectively as landfill gas: this can be burned and is a source of renewable energy. Landfill gas (LFG) can be burned either directly for heat or to generate electricity for public consumption. Landfill gas contains approximately 50 percent methane, the same gas that is found in natural gas. If landfill gas is not harvested, it escapes into the atmosphere: this is not desirable because methane is a greenhouse gas, with more global warming potential than carbon dioxide. Over a time span of 100 years, methane has a global warming potential of 23 relative to CO2. Therefore, during this time, one ton of methane produces the same greenhouse gas (GHG) effect as 23 tons of CO2. When methane burns the formula is CH4 + 2O2 = CO2 + 2H2O. So by harvesting and burning landfill gas, its global warming potential is reduced a factor of 23, in addition to providing energy for heat and power. PhD Frank Keppler and PhD Thomas Rockmann discovered that living plants also produce methane CH4. The amount of methane produced by living plants is 10 to 100 times greater than that produced by dead plants but does not increase global warming because of the carbon cycle. Anaerobic digestion can be used as a distinct waste management strategy to reduce the amount of waste sent to landfill and generate methane, or biogas. Any form of biomass can be used in anaerobic digestion and will break down to produce methane, which can be harvested and burned to generate heat, power or to power certain automotive vehicles. A 3 MW landfill power plant would power 1,900 homes. It would eliminate 6,000 tons per year of methane from getting into the environment. It would eliminate 18,000 tons per year of CO2 from fossil fuel replacement. This is the same as removing 25,000 cars from the road, or planting 36,000 acres (146 km^2) of forest, or not using 305,000 barrels of oil per year.

First Generation Biofuels

'First-generation fuels' refer to biofuels made from sugar, starch, vegetable oil, or animal fats using conventional technology. The most common first generation biofuels are listed below.

Vegetable Oil

Vegetable oil can be used for either food or fuel; the quality of the oil may be lower for fuel use. Vegetable oil can be used in many older diesel engines (equipped with indirect injection systems), but only in warm climates. In most cases, vegetable oil is used to manufacture biodiesel, which is compatible with most diesel engines when blended with conventional diesel fuel. MAN B&W Diesel, Wartsila and Deutz AG offer engines that are compatible with straight vegetable oil. Used vegetable oil is increasingly being processed into biodiesel, and at a smaller scale, cleaned of water and particulates and used as a fuel.

Biodiesel

Biodiesel is the most common biofuel in Europe. It is produced from oils or fats using transesterification and is a liquid similar in composition to mineral diesel. Its chemical name is fatty acid methyl (or ethyl) ester (FAME). Oils are mixed with sodium hydroxide and methanol (or ethanol) and the chemical reaction produces biodiesel (FAME) and glycerol. 1 part glycerol is produced for every 10 parts biodiesel.

Biodiesel can be used in any diesel engine when mixed with mineral diesel. In some countries manufacturers cover their diesel engines under warranty for 100% biodiesel use, although Volkswagen Germany, for example, asks drivers to make a telephone check with the VW environmental services department before switching to 100% biodiesel (see biodiesel use). Many people have run their vehicles on biodiesel without problems. However, the majority of vehicle manufacturers limit their recommendations to 15% biodiesel blended with mineral diesel. In many European countries, a 5% biodiesel blend is widely used and is available at thousands of gas stations.

In the USA, more than 80% of commercial trucks and city buses run on diesel. Therefore "the nascent U.S. market for biodiesel is growing at a staggering rate—from 25 million gallons per year in 2004 to 78 million gallons by the beginning of 2005. By the end of 2006 biodiesel production was estimated to increase fourfold to more than 1 billion gallons," energy expert Will Thurmond writes in an article for the July-August 2007 issue of THE FUTURIST magazine.

Bioalcohols

Biologically produced alcohols, most commonly ethanol and less commonly propanol and butanol, are produced by the action of microorganisms and enzymes through fermentation.

Butanol

Butanol is often claimed to provide a direct replacement for gasoline, because it can be used directly in a gasoline engine (in a similar way to biodiesel in diesel engines). It is not in widespread production, and engine manufacturers have not made statements about its use[verification needed]. While on paper (and a few lab tests) it appears that butanol has sufficiently similar characteristics with gasoline such that it should work without problem in any gasoline engine, no widespread experience exists. Butanol is formed by ABE fermentation (acetone, butanol, ethanol) and experimental modifications of the process show potentially high net energy gains with butanol as the only liquid product. Butanol will produce more energy and allegedly can be burned "straight" in existing gasoline engines (without modification to the engine or car), and is less corrosive and less water soluble than ethanol, and could be distributed via existing infrastructures. DuPont and BP are working together to help develop Butanol.

Bioethanol

Ethanol is the most common biofuel worldwide. This alcohol fuel is produced by fermentation of sugars derived from wheat, corn, sugar beet and sugar cane. The production methods used are enzymatic digestion (to release sugars from stored starches e.g. from wheat and corn), fermentation of the sugars, distillation and drying. Ethanol can be used in petrol engines as a replacement for gasoline; it can be mixed with gasoline to any percentage, see common ethanol fuel mixtures for information on ethanol. All petrol engines can run on blends of up to 15% bioethanol with petroleum/gasoline. For higher percentage blends, engine modifications are needed. Many car manufacturers are now producing flex-fuel vehicles, which can run on any combination of bioethanol and petrol, up to 100% bioethanol.

Methanol is currently produced from natural gas, a fossil fuel. It can also be produced from biomass (biomethanol). The methanol economy is an interesting alternative to the hydrogen economy.

BioGas

Biogas is produced by the process of anaerobic digestion of organic material by anaerobes. It can be produced either from biodegradable waste materials or by the use of energy crops fed into anaerobic digesters to supplement gas yields. The solid byproduct, digestate, can be used as a biofuel or a fertiliser.

Biogas contains methane and can be recovered from industrial anaerobic digesters and mechanical biological treatment systems. Landfill gas is a less clean form of biogas which is produced in landfills through naturally occurring anaerobic digestion. If it escapes into the atmosphere it is a potent greenhouse gas. Oils and gases can be produced from various biological wastes—Thermal depolymerization of waste can extract methane and other oils similar to petroleum. GreenFuel Technologies Corporation developed a patented bioreactor system that uses nontoxic photosynthetic algae to take in smokestacks flue gases and produce biofuels such as biodiesel, biogas and a dry fuel comparable to coal.

Solid Biofuels

Examples include wood, charcoal, and dried excrement.

Second Generation Biofuels

Second generation biofuels use biomass to liquid technology, including cellulosic biofuels from non food crops.

The following second generation biofuels are under development:

- BioHydrogen

- Bio-DME
- Biomethanol
- DMF
- HTU diesel
- Fischer-Tropsch diesel
- Mixed Alcohols (i.e., mixture of mostly ethanol, propanol and butanol, with some pentanol, hexanol, heptanol and octanol)

Bio-DME, Fischer-Tropsch, BioHydrogen diesel, Biomethanol and Mixed Alcohols all use syngas for production. This syngas is produced by gasification of biomass. HTU (High Temperature Upgrading) diesel is produced from particularly wet biomass stocks using high temperature and pressure to produce an oil.

BioHydrogen is the same as hydrogen except it is produced from a biomass feedstock. This is done using gasification of the biomass and then reforming the methane produced, or alternatively, this can be accomplished with some organisms that produce hydrogen directly under certain conditions. BioHydrogen can be used in fuel cells to produce electricity.

DMF. Recent advances in producing DMF from fructose and glucose using catalytic biomass-to-liquid process have increased its attractiveness.

Bio-DME is the same as DME but is produced from a bio-sources. Bio-DME can be produced from Biomethanol using catalytic dehydration or it can be produced from syngas using DME synthesis. DME can be used in the compression ignition engine.

Biomethanol is the same as methanol but it is produced from biomass. Biomethanol can be blended with petrol up to 10-20% without any infrastructure changes.

HTU diesel is produced from wet biomass. It can be mixed with fossil diesel in any percentage without need for infrastructure.

Fischer-Tropsch diesel (FT)diesel is produced using gas-to-liquids technology. FT diesel can be mixed with fossil diesel at any percentage without need for infrastructure change.

Mixed alcohols are produced from syngas with catalysts similar to those used for methanol. Most R&D in this area is concentrated in producing mostly ethanol. However, some fuels are marketed as mixed alcohols. Mixed alcohols are superior to pure methanol or ethanol, in that the higher alcohols have higher energy content. Also, when blending, the higher alcohols increase compatibility of gasoline and ethanol, which increases water tolerance and decreases evaporative emissions. In addition, higher alcohols have also lower heat of vaporization than ethanol, which is important for cold starts. (For another method for producing mixed alcohols from biomass see bioconversion of biomass to mixed alcohol fuels) Wood diesel A new biofuel was developed by the University of Georgia from wood chips. The oil is extracted and then added to unmodified diesel engines. Either new plants are used or planted to replace the old plants. The charcoal byproduct is put back into the soil as a fertilizer. According to the director Tom Adams since carbon is put back into the soil, this biofuel can actually be carbon negative not just carbon neutral. Carbon negative decreases carbon dioxide in the air reversing the greenhouse effect not just reducing it.

Micro Algae

Much research is being done about the use of microalgae as an energy source, with applications for biodiesel, ethanol, methanol, methane and hydrogen. The production of biofuels to replace oil and natural gas is in active development, focusing on the use of cheap organic matter (usually cellulose, agricultural and sewage waste) in the efficient production of liquid and gas biofuels which yield high net energy gain. One advantage of many biofuels over most other fuel types is that they are biodegradable, and so relatively harmless to the environment if spilled.

Biofuels in Developing Countries

Biofuel industries are becoming established in many developing countries. Many developing countries have extensive biomass resources that are becoming more valuable as demand for biomass and biofuels increases. The approaches to biofuel development in different parts of the world varies. Countries such as India and China are developing both bioethanol and biodiesel programs. India is extending plantations of jatropha, an oil-producing tree that is used in biodiesel production. The Indian sugar ethanol program sets a target of 5% bioethanol incorporation into transport fuel Ethanol India website. China is a major bioethanol producer and aims to incorporate 15% bioethanol into transport fuels by 2010. Amongst rural populations in developing countries, biomass provides the majority of fuel for heat and cooking. Wood, animal dung and crop residues are commonly burned. Figures from the International Energy Agency show that biomass energy provides around 30% of the total primary energy supply in developing countries; over 2 billion people depend on biomass fuels as their primary energy source. world resources institute document on wood fuels The use of biomass fuels for cooking indoors is a source of health problems and pollution. 1.3 million deaths were attributed to the use of biomass fuels with inadequate ventilation by the International Energy Agency in its World Energy Outlook 2006. Proposed solutions include improved stoves and alternative fuels. However, fuels are easily damaged, and alternative fuels tend to be expensive.

People in developing countries are unlikely to be able to afford to put these solutions in place. Organizations such as Intermediate Technology Development Group work to make improved facilities for biofuel use and better alternatives accessible to those who cannot get them.

Efforts and Promotion

Recognizing the importance of implementing bioenergy, there are international organizations such as IEA Bioenergy, established in 1978 by the International Energy Agency (IEA), with the aim of improving cooperation and information exchange between countries that have national programs in bioenergy research, development and deployment. In Brazil, the government hopes to build on the success of the Proálcool ethanol program by expanding the production of biodiesel which must contain 2% biodiesel by 2008, increasing to 5% by 2013. Colombia mandates the use of 10% ethanol in all gasoline sold in cities with populations exceeding 500,000. In Venezuela, the state oil company is supporting the construction of 15 sugar cane distilleries over the next five years, as the government introduces a E10 (10% ethanol) blending mandate.

An EU directive has set the goal of replacing 5.75% of transportation fuel by biofuels by 2010 in all member states. In Canada, the government aims for 45% of the country's gasoline consumption to contain 10% ethanol by 2010. In Southeast Asia, Thailand has mandated an ambitious 10% ethanol mix in gasoline starting in 2007. For similar reasons, the palm oil industry plans to supply an increasing portion of national diesel fuel requirements in Malaysia and Indonesia. In India, a bio-ethanol program calls for E5 blends throughout most of the country targeting to raise this requirement to E10 and then E20. In China, the government is making E10 blends mandatory in five provinces that account for 16% of the nation's passenger cars.

European Union

The European Union has set a goal:

- For 2010 that each member state should achieve at least 5.75% biofuel usage of all used traffic fuel.
- For 2020, 10 %.

USA

A senior member of the House Energy and Commerce Committee Congressman Fred Upton has legislation to use at least E10 fuel by 2012 in all cars in the USA. The fuel is less expensive and is cleaner for the environment. It has higher octane so all cars made after 2012 can be made with a higher compression rating to get better fuel economy.

Oregon

Oregon Governor Ted Kulongoski signed legislation in July 2007 that will require all gasoline sold in the state to be blended with 10% bioethanol (a blend known as BE10) and all diesel fuel sold in the state to be blended with 2% biodiesel (a blend known as BD2).

Current Issues in Biofuel Production and Use

Biofuels can provide benefits including: reduction of greenhouse gas emissions, reduction of fossil fuel use, increased national energy security, increased rural development and a sustainable fuel supply for the future. However, biofuels have limitations. The feedstocks for biofuel production must be replaced rapidly and biofuel production processes must be designed and implemented so as to supply the maximum amount of fuel at the cheapest cost, while providing maximum environmental benefits. Broadly speaking, first generation biofuel production processes cannot supply us with more than a few percent of our energy requirements sustainably. The reasons for this are described below. Second generation processes can supply us with more biofuel, with better environmental gains. The major barrier to the development of second generation biofuel processes is their capital cost: establishing second generation biodiesel plants has been estimated at •500million Nexant Chem Systems study

Rising Food Prices/the "Food vs. Fuel" Debate

Due to rising demand for biofuels, farmers worldwide have an increased economic incentive to grow crops for biofuel production instead of food production. Without political intervention, this could lead to reduced food production and increased food prices and inflation. The impacts of this would be greatest on poorer countries or countries that rely on imported food for their subsistence. In early 2007 there were a number of reports linking stories as diverse as food riots in Mexico due to rising prices of corn for tortillas and reduced profits at Heineken, the large international brewer, to the increasing use of corn (maize) grown in the US Midwest for bio-ethanol production. (In the case of beer, the barley area was cut in order to increase corn production. Barley is not currently used to produce bioethanol). Environmental campaigner George Monbiot has argued in the British newspaper The Guardian for a 5-year freeze on biofuels while their impact on poor communities and the environment is assessed. One problem with this approach is that economic drivers are required in order to push through the development of more sustainable second generation biofuel processes: these will be stalled if biofuel production decreases. Supporters of biofuels claim that a more viable solution is to increase political and industrial support for, and rapidity of, second generation biofuel implementation from non food crops, including cellulosic biofuels. The most recent UN report on biofuel also raises issues regarding food security and biofuel production. Jean Ziegler, the UN

Special Rapporteur on food, concluded that while the argument for biofuels in terms of energy efficiency and climate change are legitimate, the effects for the world's hungry of transforming wheat and maize crops into biofuel are "absolutely catastrophic," and terms such use of arable land a "crime against humanity." Ziegler also calls for a 5-year moratorium on biofuel production. Food surpluses exist in many developed countries. For example, the UK wheat surplus was around 2 million tonnes in 2005 (Defra figures after exports,). This surplus alone could produce sufficient bioethanol to replace around 2.5% of the UK's petroleum consumption, without requiring any increase in wheat cultivation or reduction in food supply or exports. However, above a few percent (i.e. if the UK wanted to replace more than around 5% of its fuel with biofuel), there would be direct competition between first generation biofuel production and food production. This is one reason why many view second generation biofuel production processes as increasingly important. Second generation biofuel production processes use non food crops. These include the stalks of wheat and corn, wood, special energy or biomass crops (e.g. Miscanthus) and waste biomass. These processes could utilise the waste products of current food-based agriculture to manufacture fuel sustainably. Second generation biofuel processes are in development: pilot plants are established for the production of ethanol from wheat straw and of syn-diesel from wood chippings. It is important to note that carbon in waste biomass is used by other organisms, e.g. it is broken down in the soil to produce nutrients, and provides a habitat for wildlife. The large scale use of such "waste" biomass by humans might threaten these habitats and organisms.

Biofuel Prices

Retail, at the pump prices, including Federal and state motor taxes, B2/B5 prices for low-level Biodiesel (B2-B5) are lower than petroleum diesel by about 12 cents, and B20 blends are the same as petrodiesel.

Poverty Reduction

Researchers at the Overseas Development Institute have argued that biofuels could help to reduce poverty in the developing world, through increased employment, wider economic growth multipliers and energy price effects. However, this potential is described as 'fragile', and is reduced where feedstock production tends to be large scale, or causes pressure on land access. With regards to potential for poverty reduction, biofuels rely on many of the same policy, regulatory or investment shortcomings that impede agriculture as a route to poverty reduction. As many of these shortcomings require policy improvements at a country level, rather than a global one, they argue for a country-by-country analysis of the potential poverty impacts of biofuels. This would consider, among other things, land administration systems, market coordination and prioritising investment in biodiesel as this 'generates more labour, has lower transportation costs and uses simpler technology'.

Energy Efficiency and Energy Balance of Biofuels

Production of biofuels from raw materials requires energy (for farming, transport and conversion to final product as well as the production of fertilizers, pesticides and herbicides). The level of energy expenditure varies by location: more intensive agricultural regimes such as those found in Western countries are more energy intensive. The more machinery is used for farming, the greater the energy expended in the process; developing countries tend to have less intensive agricultural methods. It is possible to produce biomass without incurring large agricultural energy costs: for example, wild-harvesting excess wood from established forests can be done without much energy input. However the yield of biomass from such resources is not consistent or large enough to support biofuel manufacture on a large scale. The energy balance of a biofuel is determined by the amount of energy put into the manufacture of fuel compared to the amount of energy released when it is burned in a vehicle and some biofuels can produce up to 2-36 times the input rate of fossil fuels. Biofuels tend to require higher energy inputs per unit energy than fossil fuels: oil can be pumped out of the ground and processed more efficiently than biofuels can be grown and processed. However, this is not necessarily a reason to use oil instead of biofuels, nor does it have an impact on the environmental benefits provided by a given biofuel. Other factors connected to energy balance are a) cost and b) environmental impact. High energy impacts do not necessarily mean that the resulting fuel will be bad for the environment: energy can be derived from renewable resources to power biofuel manufacture. Energy balance is not necessarily a measure of a good biofuel. Biofuels should be affordable, sustainable, abundant and provide good GHG emissions savings when compared with fossil fuels. Energy balance/ efficiency of conversion is relevant when considering how best to use a given amount of biomass resources. For example, given limited resources should biomass be converted into heat and power or liquid transport fuels? Looking at energy balance and the efficiency of energy conversion can help to use biomass resources efficiently and with maximum environmental gain. Studies have been done that calculate energy balances for biofuel production. Some of these show large differences depending on the biomass feedstock used and location. The energy balance is more favourable for biofuels made from crops grown in subtropical or tropical areas than those made from crops grown in temperate areas. This is largely due to the increased yield of biomass from crops in areas that receive more sunlight.

Life cycle assessments of biofuel production show that under certain circumstances, biofuels produce only limited savings in energy and greenhouse gas emissions. Fertiliser inputs and transportation of biomass across large distances can reduce the GHG savings achieved. The location of biofuel processing plants can be planned to minimize the need for transport, and agricultural regimes can be developed to limit the amount of fertiliser used for biomass production. A European study on the greenhouse gas emissions found that well-to-wheel (WTW) CO2 emissions of biodiesel from seed crops such as rapeseed could be almost as high as fossil diesel. It showed a similar result for bio-ethanol from starch crops, which could have almost as many WTW CO2 emissions as fossil petrol. This study showed that second generation biofuels have far lower WTW CO2 emissions. Other independent LCA studies show that biofuels save around 50% of the CO2 emissions of the equivalent fossil fuels. This can be increased to 80-90% GHG emissions savings if second generation processes or reduced fertiliser growing regimes are used (Concawe Well to Wheels LCA for biofuels).

Environmental Effects

Some mainstream environmental groups support biofuels as a significant step toward slowing or stopping global climate change. However, biofuel production can threaten the environment if it is not done sustainably. This finding has been backed by reports of the UN, the IPCC and some other smaller environmental and social groups as the EEB and the Bank Sarasin, which generally remain negative about biofuels. As a result, governmental and environmental organisations are turning against biofuels made at a non-sustainable way (hereby preferring certain oil sources as jatropha and lignocellulose over palm oil) and are asking for global support for this. Also, besides supporting these more sustainable biofuels, environmental organisations are redirecting to new technologies that do not use combustion engines as hydrogen and compressed air.

Biofuels produce greenhouse gas emissions during their manufacture. The source of these emissions are: fertilisers and agricultural processing, transportation of the biomass, processing of the fuels, and transport and delivery of biofuels to the consumer. Some biofuel production processes produce far fewer emissions than others; for example sugar cane cultivation requires fewer fertiliser inputs than corn cultivation, therefore sugar cane bioethanol reduces greenhouse gas emissions more effectively than corn derived bioethanol. However, given the appropriate agricultural techniques and processing strategies, biofuels can provide emissions savings of at least 50% when compared to fossil fuels such as diesel and petroleum. The increased manufacture of biofuels will require increasing land areas to be used for agriculture. Second generation biofuel processes can ease the pressure on land, because they can use waste biomass, and existing (untapped) sources of biomass such as crop residues and potentially even marine algae.

In some regions of the world, a combination of increasing demand for food, and increasing demand for biofuel, is causing deforestation and threats to biodiversity. The best reported example of this is the expansion of oil palm plantations in Malaysia and Indonesia, where rainforest is being destroyed to establish new oil palm plantations. It is an important fact that 90% of the palm oil produced in Malaysia is used by the food industry Malaysian Palm Oil Council; therefore biofuels cannot be held solely responsible for this deforestation. There is a pressing need for sustainable palm oil production for the food and fuel industries; palm oil is used in a wide variety of food products. The Roundtable on Sustainable Biofuels is working to define criteria, standards and processes to promote sustainably produced biofuels. Palm oil is also used in the manufacture of detergents, and in electricity and heat generation both in Asia and around the world (the UK burns palm oil in coal-fired power stations to generate electricity). Significant area is likely to be dedicated to sugar cane in future years as demand for ethanol increases worldwide. The expansion of sugar cane plantations will place pressure on environmentally-sensitive native ecosystems including rainforest in South America. In forest ecosystems, these effects themselves will undermine the climate benefits of alternative fuels, in addition to representing a major threat to global biodiversity. Although biofuels are generally considered to improve net carbon output, biodiesel and other fuels do produce local air pollution, including nitrogen oxides, the principal cause of smog.

Glossary

Antibody: A protein produced by certain white blood cells in response to a foreign substance (antigen). Each antibody can bind only to a specific antigen.

Antigen: Any foreign or "non-self" substance that, when introduced into the body, causes the immune system to create an antibody.

Base: Alkaline chemical substance, in particular the cyclic nitrogen compounds found in DNA and RNA.

Base-pairing: When two complementary bases (A with T or G with C) recognize each other and are held together by hydrogen bonds.

Biotechnology: A set of biological techniques developed through basic research and now applied to research and product development.

Biotin: A colorless crystalline vitamin, C10H16N2O3S, of the vitamin B complex, essential for the activity of many enzyme systems and found in large quantities in liver, egg yolk, milk, and yeast. Also known as vitamin H.

Cell: Small, watery, membrane-bound compartment filled with chemicals; the basic unit of any living thing.

Clone: A group of identical genes, cells, or organisms derived from a single ancestor.

Denature: To destroy the 3-D folding of a protein or other polymeric molecule.

Deoxyribonuclease: An enzyme that cuts DNA into shorter pieces.

Deoxyribose: The sugar with five carbon atoms that is found in DNA.

DNA: The substance of heredity; a linear molecule that carries the genetic information that cells need to replicate and to produce proteins and DNA. DNA stands for deoxyribonucleic acid.

DNA Cloning: Facilitates isolation and manipulation of fragments of an organism's genome by replicating them as part of an independent vector.

DNA Double Helix: Most common form of DNA. Two separate and antiparallel chains of DNA are wound around each other in a right - handed helical path, with sugar - phosphate backbones on the outside and bases connected by hydrogen bonds on the inside. A (Adenine) pairs with T (Thymine); G (Guanine) pairs with C (Cytosine). The two chains are complementary.

DNA Libraries: Consist of sets of random cloned fragments of either genomic or cDNA, each in a separate vector molecule. Are used in isolation of unknown genes.

DNA Sequencing: The widely used Sanger's enzymic method uses dideoxynucleotide's as chain terminators to produce a ladder of molecule's generated by polymerase extension of a primer. Determines the sequences of the bases in DNA.

Double-helix: Structure in which two strands of DNA are twisted spirally around each other.

Electrophoresis: Movement of charged molecules towards an electrode of the opposite charge; used to separate nucleic acids and proteins.

Enzyme: A protein that acts as a catalyst, affecting the rate at which chemical reactions occur in cells.

Expression of a gene: Synthesis of the protein, or the gene product - result of transcription plus processing plus translation.

Gel electrophoresis: Electrophoresis of charged moleculaes thorugh a gel meshwork in order to sort them by size.

Gene: A unit of inheritance; a working subunit of DNA. Each of the body's 50,000 to 100,000 genes contains the code for a specific product, typically, a protein such as an gene's coded information is translated into the structures present and operating in the cell (either proteins or RNAs).

Gene therapy: Treatment that alters genes - the basic units of heredity found in all cells in the body.

Genetic code: The sequence of nucleotides, coded in triplets (codons) along the mRNA, that determines the sequence of amino acids in protein synthesis.

Genome: All the genetic material in the chromosomes of a particular organism; its size is generally given as its total number of base pairs.

Genomic DNA library: Library made from chromosomal DNa which therefore contains introns.

Genotype: Genetic constitution of an organism; describes an organism at the genetic leve.

Intron: Segment of a gene that does not code for protein.

Mitosis: The process of nuclear division in cells that produces daughter cells that are genetically identical to each other and to the parent cell.

Nucleotide: A molecule consisting of a nucleoside (purine or pyrimidine base plus a pentose sugar)with one or more phosphate groups; a monomer or subunit of nucleic acid, consisting of a base plus sugar plus 3 phosphates.

Nucleus: The nucleus of a cell is an internal compartment surrounded by the nuclear membrane and containing the chromosomes. Only the cells of higher organisms have nuclei.

Organelle: A discrete subcellular structure that has a specialised function (eg, nucleus, mitochondrion).

Pathway elucidcation: The term biochemical pathways has principally referred to metabolic pathways, which are the pathways by which a cell converts compounds that enter it into cellular components (*e.g.*, small molecules and macromolecules including proteins, nucleic acids, storage carbohydrates, and fatty acids) and by which the cell derives energy. Signaling pathways are biochemical pathways that regulate cellular characteristics and processes such as physiology, proliferation, changes in shape and motility, differentiation, adhesion, and intercellular interactions. Pathway elucidation refers to the understanding of these.

Peptide: A molecule consisting of 2 to approximately 20 amino acids connected by peptide bonds; a short segment of a larger protein or a completely functional molecule unto itself.

Protein: A large, complex molecule composed of amino acids. Wide variety of functions *e.g.* enzymes catalysing biochemical reactions, membrane receptors in cell signalling, transport and storage, antibody proteins, nutritional proteins, contractile muscle fibers (actin and myosin) etc.

Proteomics: The identification and analysis of the total protein complement expressed by any given cell type under defined conditions.

Receptor: A protein or group of associated proteins in a cell or on its surface that selectively binds a specific substance (called a ligand).

Recombination: The natural process of breaking and rejoining DNA strands to produce new combinations of genes and, thus, generate genetic variation.

RNA: A chemical found in the nucleus and cytoplasm of cells; it plays an important role in protein synthesis and other chemical activities of the cells from which all blood cells develop.

Telomere: The ends of chromosomes. These specialized structures are involved in the replication and stability of linear DNA molecules.

Transcription: The process of copying information from DNA into new strands of messenger RNA (mRNA).

Translation: The process of turning instructions from mRNA, base by base, into chains of amino acids that then fold into proteins.

Umbilical cord: The structure that connects the placenta and the embryo; contains the umbilical arteries and the umbilical vein.

Unicellular: Single-celled.

Uniformitarianism: The idea that geological processes have remained uniform over time and that slight changes over long periods can have large-scale consequences; proposed by James Hutton in 1795 and reÞned by Charles Lyell during the 1800s. The principle on which modern geology was founded: processes operating today on the earth operated in much the same way in the geologic past. Sometimes expressed as "the present is the key to the past".

Uninucleate: Term applied to cells having only a single nucleus.

Unsaturated fat: A triglyceride with double coavent bonds between some carbon atoms.

Uracil: The pyrimidine that replaces thymine in RNA molecules and nucleotides.

Ureter: A muscular tube that transports urine by peristaltic contractions from the kidney to the bladder.

Urethra: A narrow tube that transports urine from the bladder to the outside of the body. In males, it also conducts sperm and semen to the outside.

Urine: Fluid containing various wastes that is produced in the kidney and excreted from the bladder.

Uterus: The organ that houses and nourishes the developing embryo and fetus. The womb. Female reproductive organ in which the fertilized egg implants.

Vaccination: The process of protecting against infectious disease by introducing into the body a vaccine that stimulates a primary immune response and the production of memory cells against the disease-causing agent.

Vaccine: A preparation containing dead or weakened pathogens that when injected into the body elicit an immune response.

Vacuoles: Membrane-bound þuid-Þlled spaces in plant and animal cells that remove waste products and store ingested food.

Vagina: The tubular organ that is the site of sperm deposition and also serves as the birth canal.

Vascular bundle: Groups of xylem, phloem and cambium cells in stems of plants descended from the procambium embryonic tissue layer.

Vascular cambium: A layer of lateral meristematic tissue between the xylem and phloem in the stems of woody plants. Lateral meristem tissue in plants that produces secondary growth.

Vascular cylinder: A central column formed by the vascular tissue of a plant root; surrounded by parenchymal ground tissue.

Vascular parenchyma: Specialized parenchyma cells in the phloem of plants.

Vascular plants: Group of plants having lignified conducting tissue (xylem vessels or tracheids).

Vascular system: Specialized tissues for transporting þuids and nutrients in plants; also plays a role in supporting the plant; one of the four main tissue systems in plants.

Vas deferens: The duct that carries sperm from the epididymis to the ejaculatory duct and urethra. The tube connecting the testes with the urethra.

Ventricle: The chamber of the heart that pumps the blood into the blood vessels that carry it away from the heart. The lower chamber of the heart through which blood leaves the heart.

Venules: The smallest veins. Blood þows into them from the capillary beds. Small veins that connect a vein with capillaries.

Vernalization: ArtiÞcial exposure of seeds or seedlings to cold to enable the plant to þower.

Vertebrae: The segments of the spinal column; separated by disks made of connective tissue (sing.: vertebra).

Vertebrate: Any animal having a segmented vertebral column; members of the subphylum Vertebrata; include reptiles, Þshes, mammals, and birds.

Vesicles: Small membrane-bound spaces in most plant and animal cells that transport macromolecules into and out of the cell and carry materials between organelles in the cell.

Vessel elements: Short, wide cells arranged end to end, forming a system of tubes in the xylem that moves water and solutes from the roots to the rest of the plant. Large diameter cells of the xylem that are extremely specialized and efficient at conduction. An evolutionary advance over tracheids. Most angiosperms have vessels.

Vestigial structures: Nonfunctional remains of organs that were functional in ancestral species and may still be functional in related species; e.g., the dewclaws of dogs.

Villi: Finger-like projections of the lining of the small intestine that increase the surface area available for absorption. Also, projections of the chorion that extend into cavities Þlled with maternal blood and allow the exchange of nutrients between the maternal and embryonic circulations. Projections of the inner layer of the small intestine that increase the surface area for absorbtion of food.

Viroids: Infective forms of nucleic acid without a protective coat of protein; unencapsulated single-stranded RNA molecules. Naked RNA, possibly of degenerated virus, that infects plants.

Virus: Infectious chemical agent composed of a nucleic acid (DNA or RNA) inside a protein coat.

Vitamins: A diverse group of organic molecules that are required for metabolic reactions and generally cannot be synthesized in the body.

Vulva: A collective term for the external genitals in women.

White blood cell: Component of the blood that functions in the immune system. Also known as a leukocyte.

Xerophytic leaves: The leaves of plants that grow under arid conditions with low levels of soil and water. Usually characterized by water-conserving features such as thick cuticle and sunken stomatal pits.

X-ray diffraction: Technique utilized to study atomic structure of crystalline substances by noting the patterns produced by x-rays shot through the crystal.

Xylem: Tissue in the vascular system of plants that moves water and dissolved nutrients from the roots to the leaves; composed of various cell types including tracheids and vessel elements. Plant tissue type that conducts water and nutrients from the roots to the leaves.

Z lines: Dense areas in myoÞbrils that mark the beginning of the sarcomeres. The actin Þlaments of the sarcomeres are anchored in the Z lines.

Zone of physiological stress The area in a population's geographic range where members of population are rare due to physical and biological limiting factors.

Zygospore: In fungi, a structure that forms from the diploid zygote created by the fusion of haploid hyphae of different mating types. After a period of dormancy, the zygospore forms sporangia, where meiosis occurs and spores form.

Zygote: A fertilized egg. A diploid cell resulting from fertilization of an egg by a sperm cell.

Bibliography

Ahmadian, A., B. Gharizadeh, A.C. Gustafsson, F. Sterky, P. Nyren, M. Uhlen, and J. Lundeberg (2000). Single-nucleotide polymorphism analysis by pyrosequencing. *Anal Biochem*, 280(1): p. 103-10.

Begon, Michael; Townsend, CR & Harper, JL (2005). *Ecology: From Individuals to Ecosystems*, 4th edition, Blackwell Publishing Limited.

Brennan, Mark G. and Mark A. Tooley (2000). "Ethics and the Biomedical Engineer". *Engineering Science and Education Journal* 9.1, 5-7.

Casada, Mark E. and James A. DeShazer (1995). "Teaching Professionalism, Design, & Communications to Engineering Freshmen". *ASEE Annual Conference Proceedings*, v 1, "Investing in the Future",1381-1386.

Cook, S.B., Russ, W.T., and Goodfred, D.W. (2005). Emory River Watershed biological assessment (pdf file). Quarterly Report submitted to the U.S. Fish and Wildlife Service and Tennessee Wildlife Resources Agency.

Cook, S.B., Russ, W.T., and Goodfred, D.W. (2006). *Biological survey of the Emory River Watershed, Tennessee* (pdf file). Final Report submitted to the U.S. Fish and Wildlife Service and Tennessee Wildlife Resources Agency.

Cristianini, N. and Hahn, M. (2006). *Introduction to Computational Genomics*, Cambridge University Press.

Curtis, H. and Barnes, N. (1989). Biology. New York: Worth Publishers.

Friedberg, E.C. (1985). *DNA Repair*. New York: WH Freeman and Company.

Fruton, J.S. (1999). *Proteins, Enzymes, Genes: The Interplay of Chemistry and Biology*. Yale University Press.

Graslund, T., M. Ehn, G. Lundin, M. Hedhammar, M. Uhlen, P.A. Nygren, and S. Hober (2002). Strategy for highly selective ion-exchange capture using a charge-polarizing fusion partner. *J Chromatogr A*, 942(1-2): p. 157-66.

Graslund, T., M. Hedhammar, M. Uhlen, P.A. Nygren, and S. Hober (2002). Integrated strategy for selective expanded bed ion-exchange adsorption and site-specific protein processing using gene fusion technology. *J Biotechnol*, 96(1): p. 93-102.

Larsson, M., S. Stahl, M. Uhlen, and A. Wennborg (2000). Expression profile viewer (ExProView): a software tool for transcriptome analysis. *Genomics*, 63(3): p. 341-53.

Lerner M, Corcoran M, Cepeda D, Nielsen ML, Zubarev R, Ponten F, Uhlen M, Hober S, Grander D, Sangfelt O. (2007). The RBCC Gene RFP2 (Leu5) Encodes a Novel Transmembrane E3 Ubiquitin Ligase Involved in ERAD. *Mol Biol Cell*. 2007 Feb 21; [Epub ahead of print]

Magner, LN (2002). *A History of the Life Sciences*. TF-CRC.

Mulcahy, Henry (1989). "A Simulated Laboratory Exercise in Genetic Engineering." Abstracts of the Annual Meeting of American Society for Microbiologists.

Mulcahy, Henry (1989). "A Simulated Laboratory Exercise in Genetic Engineering." Abstracts of the Annual Meeting of American Society for Microbiologists.

National Park Service. (2005). *Obed National Wild and Scenic River* . Retrieved September 11, 2006, from http://www2.nature.nps.gov/geology/parks/obri/index.cfm

Naurato, N. and T.J. Smith (2003). "Ethical Considerations in Bioengineering Research". *Biomedical Sciences Instrumentation* 39: 573-8.

Nord, K., O. Nord, M. Uhlen, B. Kelley, C. Ljungqvist, and P.A. Nygren (2001). Recombinant human factor VIII-specific affinity ligands selected from phage-displayed combinatorial libraries of protein A. *Eur J Biochem*, 268(15): p. 4269-77.

Nord, O., M. Uhlen, and P.A. Nygren (2003). Microbead display of proteins by cell-free expression of anchored DNA. *J Biotechnol*, 106(1): p. 1-13.

Pienkowski, D. (2000). "The Need for a Professional Code of Ethics in Biomedical Engineering: A Lesson from History". *Critical Reviews in Biomedical Engineering* 28.3-4: 513-6.

Retzer, W.J. (1991). *Biotechnology Workbook.* Englewood Cliffs, NJ: Prentice Hall.

Ronaghi, M., M. Uhlen, and P. Nyren (1998).A sequencing method based on real-time pyrophosphate. *Science*, 281(5375): p. 363, 365.

Ronnmark, J., C. Kampf, A. Asplund, I. Hoiden-Guthenberg, K. Wester, F. Ponten, M. Uhlen, and P.A. Nygren (2003). Affibody-beta-galactosidase immunoconjugates produced as soluble fusion proteins in the Escherichia coli cytosol. *J Immunol Methods*, 281(1-2): p. 149-60.

Ronnmark, J., H. Gronlund, M. Uhlen, and P.A. Nygren (2002). Human immunoglobulin A (IgA)-specific ligands from combinatorial engineering of protein A. *Eur J Biochem*, 269(11): p. 2647-55.

Ronnmark, J., M. Hansson, T. Nguyen, M. Uhlen, A. Robert, S. Stahl, and P.A. Nygren

(2002). Construction and characterization of affibody-Fc chimeras produced in Escherichia coli. *J Immunol Methods*, 261(1-2): p. 199-211.

Russ, W.T. (2006). *Current distribution and seasonal habitat use of the threatened spotfin chub in the Emory River Watershed* (pdf file). Master's Thesis. Tennessee Technological University: Cookeville.

Saha, Atanu; Grabowski, Henry; Birnbaum, Howard; Greenberg, Paul; Bizan, Oded (2006). "Generic Competition in the U.S. Pharmaceutical Industry," *International Journal of the Economics of Business*, 13(1): 15-38(24)

Sasser, M.A. (2003). *Indicator bacteria levels in the waters of two east Tennessee watersheds* . Master's Thesis. University of Tennessee: Knoxville.

Schellekens H. (2004). "Biosimilar Therapeutic Agents: Issues with Bioequivalence and Immunogenicity," *Eur J Clin Invest.* 34(12):797-799.

Schellekens, Huub (2002). "Bioequivalence and the Immunogenicity of Biopharmaceuticals," *Nature Reviews*; 1: 457-462

Schmalzer, P.A. (1982). *Vegetation of the Obed Wild and Scenic River, Tennessee, and a comparison of reciprocal averaging ordination and binary discriminant analysis* Doctoral Dissertation. University of Tennessee: Knoxville.

Schwartz, Lewis B. (2002). "The Dilemma of Bioengineering Research on Human Subjects". Proceedings of the 2002 IEEE Engineering in Medicine and Biology 24th Annual Conference and the 2002 Fall Meeting of the Biomedical Engineering Society (BMES / EMBS), Oct 23-26 2002, Houston, TX. *Annual International Conference of the IEEE Engineering in Medicine and Biology - Proceedings* v. 3: 2668-2669.

Schwenk JM, Lindberg J, Sundberg M, Uhlen M, Nilsson P. (2007). Determination of Binding Specificities in Highly Multiplexed Bead-based Assays for Antibody Proteomics. *Mol Cell Proteomics*. 6(1):125-132.

Slater, R.J. et al. (1986). *Experiments in Molecular Biology*. Clifton, NJ: Humana Press.

Smith, G.L., Mackett, M., and Ross, B. (1983). "Infectious vaccinia virus recombinants that express hepatitis B virus surface antigen." Nature . 302:490-495. Therman, E. and Susman, M. (1993) Human Chromosomes: Structures, Behavior and Effects 3rd Edition. New York: Springer-Verlag.

Smith, R.P. (1979). *Status report on the occurrence of the spotfin chub,* Hyboposis monacha (Cope), *in the Emory River, Morgan County, Tennessee* (pdf file). Gatlinburg: Fishery Resources.

Spath, P.J., A.G. Sjoholm, G.N. Fredrikson, G. Misiano, R. Scherz, U.B. Schaad, B. Uhring-Lambert, G. Hauptmann, J. Westberg, M. Uhlen, C. Wadelius, and L.

Truedsson (1999). Properdin deficiency in a large Swiss family: identification of a stop codon in the properdin gene, and association of meningococcal disease with lack of the IgG2 allotype marker G2m(n). *Clin Exp Immunol*, 118(2): p. 278-84.

Speir, H.J. (1969). *Comparison of rock bass,* Ambloplites rupestris (Rafinesque) *populations in Obed River and Spring Creek, Tennessee* . Master's Thesis. Tennessee Technological University: Cookeville.

Stedman, S., and Stedman, B. (2002). *Notes on the birds of the Big South Fork National River and Recreation Area and the Obed River National Wild and Scenic River* . Cookeville: Tennessee Technological University.

Stenvall M, Steen J, Uhlen M, Hober S, Ottosson J. (2005). "High-throughput solubility assay for purified recombinant protein immunogens" *Biochim Biophys Acta.*, 31;1752(1):6-10

Sterky F, Bhalerao RR, Unneberg P, Segerman B, Nilsson P, Brunner AM, Charbonnel-Campaa L, Lindvall JJ, Tandre K, Strauss SH, Sundberg B, Gustafsson P, Uhlen M, Bhalerao RP, Nilsson O, Sandberg G, Karlsson J, Lundeberg J, Jansson S. (2004). A Populus EST resource for plant functional genomics. *Proc Natl Acad Sci U S A.* 101(38):13951-6.

Vener, T., M. Nygren, A. Andersson, M. Uhlen, J. Albert, and J. Lundeberg (1998). Use of multiple competitors for quantification of human immunodeficiency virus type 1 RNA in plasma. *J Clin Microbiol*, 36(7): p. 1864-70.

Waessmann, H., S. Zollner, A.C. Gustafsson, V. Wiebe, M. Laan, J. Lundeberg, M. Uhlen, and S. Paabo (2002). Extensive linkage disequilibrium in small human populations in Eurasia. *Am J Hum Genet*, 70(3): p. 673-85.

Index

D

E

F

G

H

I